P9-CCL-153
Polyester:
The Indestructible Fashion
TIFFANY
CITY
LIGHTING
Schiffer Publishing Ltd
4880 Lower Valley Rd. Atglen, PA 19310 USA
Matthew Boyd Smith

This work is dedicated to my parents,
Bruce and Desire Smith

Library of Congress Catalog Card Number: 98-84543

Designed by Bonnie M. Hensley
Layout by Randy L. Hensley
Typeset in Van Dijk bold/Swiss 721 BT

ISBN: 0-7643-0424-0
Printed in China
1 2 3 4

Published by Schiffer Publishing Ltd.
4880 Lower Valley Road
Atglen, PA 19310
Phone: (610) 593-1777; Fax: (610) 593-2002
E-mail: Schifferbk@aol.com

In Europe, Schiffer books are distributed by Bushwood Books
6 Marksbury Avenue Kew Gardens
Surrey TW9 4JF England
Phone: 44(0)181-392-8585; Fax: 44(0) 181-392-9876
E-mail: Bushwd@aol.com

Contents

Acknowledgments

My sincere thanks goes to Peter and Nancy Schiffer for their faith in polyester—and in me! Thanks also go to fashion models Beth Geronikos, Peter N. Schiffer, Katherine Wellington, and Kayla Agran and to the *Black Banana* and *Polly Esthers* for permitting me to produce and photograph fashion shows at their clubs. Thanks go to my brother, Michael Smith, for assisting with the photo shoots. Thanks go to David Sterner for permission to use his blue and red jumpsuit, pictured on page 136. Also, my thanks go to Jane Lipton at the *Antiques Marketplace* in Manayunk, Philadelphia, Pennsylvania, for believing in the marketability of polyester, and to Anita Buonassisi at *Amarcord* in Philadelphia for her faith in my abilities as a designer and fashion trend forecaster. Special thanks goes to Beth Geronikos for her inspirational belief in my abilities and future successes, and her creative energy helping me.

Matthew Boyd Smith enjoying himself after successfully gathering all the information and material needed for this book.

Introduction: What is It?

It's indestructible. It won't shrink, won't wrinkle, is color fast, and the moths can't get to it. It's polyester!

Invented in 1941 in Lancaster, England, by J. T. Dickson and J. R. Whinfield, polyester fiber was first manufactured in England by Calico Prints (J. Gordon Cook, *Handbook of Textile Fibers*, 1984). Polyester is the generic name for the man-made fibers developed from ethylene glycol and terephthalic acid (*Fairchild's Dictionary of Fashion*, 2nd Edition). The manufacturing of polyester was suspended during the Second World War but the rights to manufacture polyester in the United States were bought by the chemical manufacturer E. I. du Pont de Nemours of Wilmington, Delaware, in 1947. Considered an experimental fiber, polyester was first named "Fiber V," and was not actually manufactured by the du Pont company until 1950. The rest is history!

Trademarks

In the early 1950s, Dacron fabrics became popular for new clothing designs. Dacron is the trademark owned by du Pont for approximately seventy types of polyester filament yarns, staple yarns, fabrics, and fiberfill— all given a type number and made for various end uses. Polyester was combined with cotton for men's shirts and with wool for coats and suits. Another popular polyester trade name is Fortrel, the trademark for a polyester fiber jointly owned by Imperial Chemical Industries Ltd. and Celanese Corporation. Equally popular is the trade name Kodel, a polyester fiber trademarked by Eastman Kodak Company.

Versatility

Polyester is pervasive and versatile. Since its introduction, polyester has grown dramatically in popularity. It could be woven with crepe texture or as a double knit. The ability of polyester fibers to take on bright colors gave it added popularity, and by the late 1960s polyester began to appear in electrifying shades—bright and vibrating hues that demanded to be noticed. What started as "broadcloth of polyester and cotton" became, two decades later, the smooth and care-free fabric of choice for the "disco generation" of the 1970s.

Printed Art

By the early 1970s, the printed designs for polyester fabric had become art. The print design motifs are so diverse that it is difficult to categorize them. Some are related to the great twentieth century art movements: Cubism, Surrealism, and Expressionism. Others show the influences of nature: wildlife, clouds, trees, and water. Still other print designs recall man's exploration of outer-space and his participation in athletics. Many designs are asymmetrical, showing large objects next to small objects, or persons next to groups of people. City themes, such as subway scenes and skyscrapers, are encountered. Even the solid color fabrics are vibrant, very often being stitched at the pockets and lapels with threads of contrasting colors.

Fashion to Last

By combining this versatile fabric with original and provocative print designs into new fashion styles, clothing manufacturers made this decade truly memorable for fashion. And the clothing is highly sought after today because it is *indestructible*. Popular styles include shirts with tight-fitting sleeves and turned-down collars, pants with flared legs or bell-bottoms, jackets with wide lapels, dresses of all lengths and descriptions, mini-skirts, vests, leisure suits, and jumpsuits.

Polyester: The Indestructible Fashion is a picture book which explores the art of polyester prints in collectible and wearable polyester clothing from the 1970s. Creative photographs taken in urban settings present this dynamic clothing that projects energy of its own. Today, in the late 1990s, the younger generation has embraced polyester once again! Not only are today's fashion designers clamoring to redesign the styles from the past, but the market is growing, too, for the exciting, one-of-a-kind, vintage pieces. It is truly a style of fashion that will not fade.

Price Reference

The price reference is intended to give collectors and dealers a general idea of what they can expect to pay for the garments pictured, in excellent condition. Price ranges are included in the text, after the descriptive captions. Highest prices are paid for unusual fabric designs. At this time, polyester is relatively inexpensive to collect, but considering the uniqueness of the fabric, and the artistic nature of its design, it will very likely escalate in price quickly.

Chapter One
Dresses & Gowns

Floral Designs

Fabric Designs

In the polyester clothing displayed in this chapter, the styles of the dresses and gowns are less important than the fabric designs. The fabrics are so colorful and so intense that the tailoring becomes somewhat less significant, but the styles have details that should be noticed.

Styling

Polyester drapes well, and most of the styles reflect this quality. Many of the dresses shown have long-sleeves and fairly straight skirts—to show the print designs to better advantage. Jackets are short and some are tied at the waist. The coat dress was a popular style in the 1970s. Halter-top gowns, sometimes with matching jackets, also were popular. It is not uncommon to find a dress with a bodice and skirt in contrasting colors, or a floral or striped top with a solid skirt. Turtlenecks and Mandarin collars are not unusual. Slits at the sides of long skirts, detailing and stitching in white or a lighter color than the garment, same-fabric sashes, front-zippers, and plain buttons are commonplace.

Yellow, orange, blue, and black "shells and flowers" print dress with a turtleneck. Labeled: 100% polyester. ($65-70)

Lilac, yellow, blue, and black Oriental designs on a coral ground; gown with matching jacket. ($75-85)

A "tulip garden" dress; red tulips on a black ground. Labeled: a Honeycomb dress, 100% polyester. ($85-95)

Brown, black, yellow, gray, and white floral print dress. ($50-55)

Red, black, and white bold floral print sleeveless dress. ($40-45)

A black mini-dress with blue and red roses. ($50-55)

A flower garden gown in shades of rose, pink, and blue on a black ground. Labeled: a Brief Originals dress, 100% polyester. ($65-70)

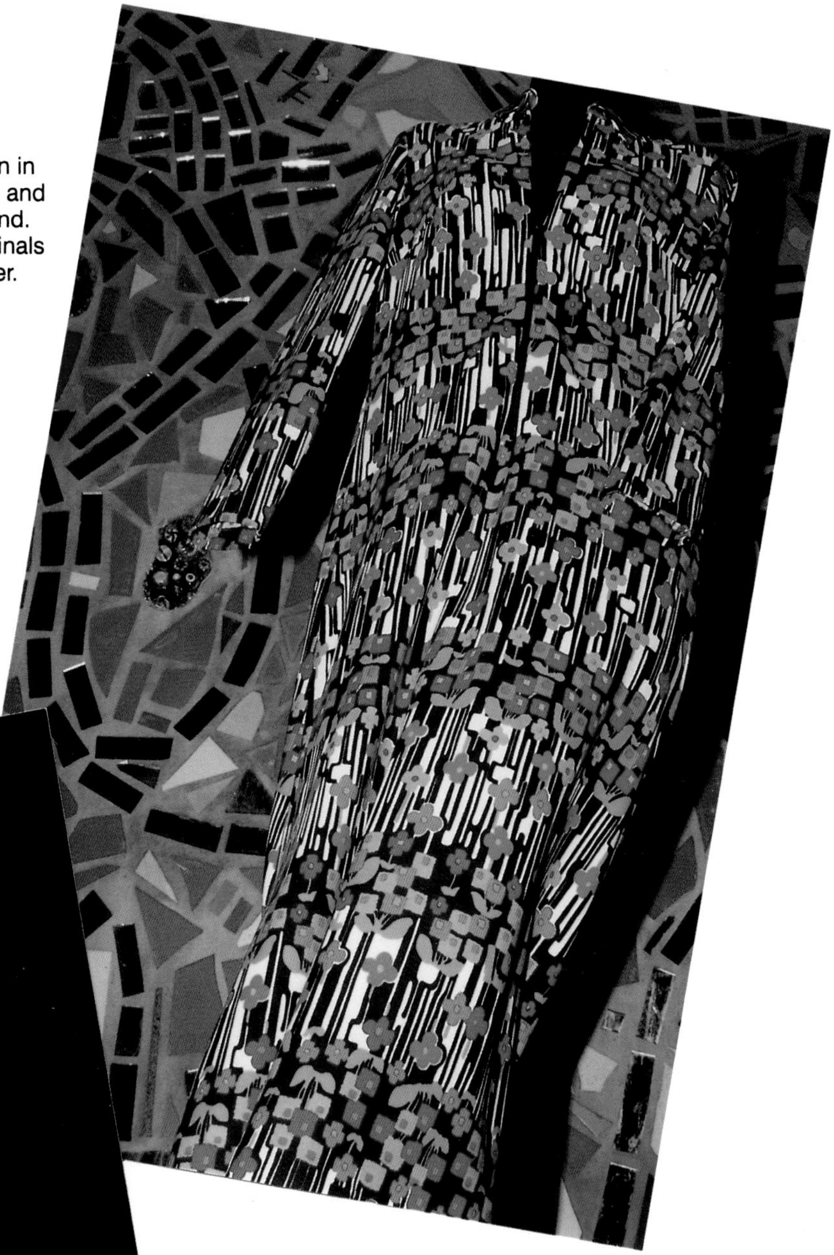

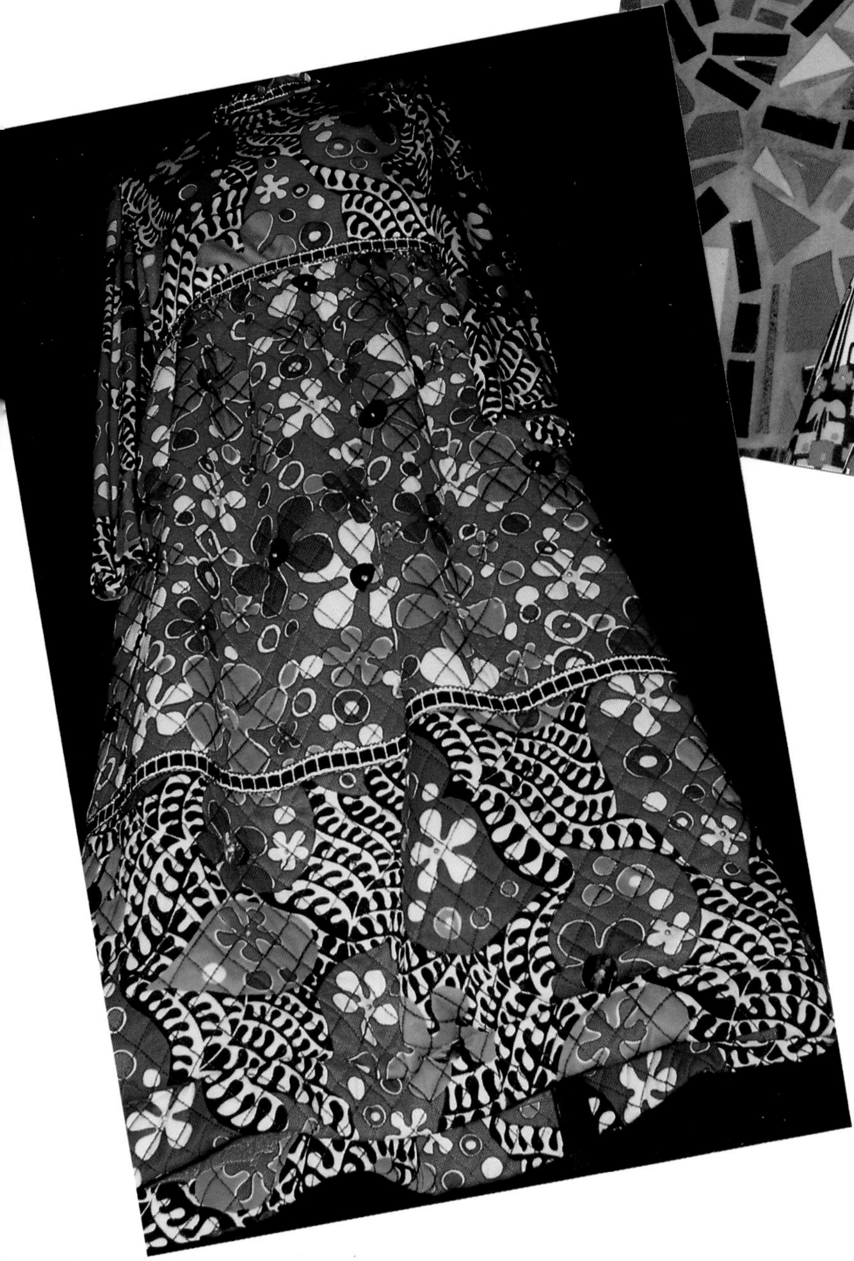

Red, green, white, purple, and black geometric floral print quilted gown. ($65-70)

Pink, white, and green "lilies of the valley" print sleeveless dress. Labeled: Dorothy Bullitt (*Philadelphia retailer, no longer in business*). ($30-35)

Maroon and beige floral print long-sleeved dress. ($25-30)

Coral, white, and beige floral print long-sleeved dress. Labeled: Leslie Fay. ($30-35)

Long-sleeved, zipper-front robe with a creme, green, and brown "Safari" print. ($30-35)

Gold, white, and pink "stripes and flowers" long-sleeved hostess gown. Labeled: Allegro. ($40-45)

Rust, brown, creme, black, and aqua "mushrooms" print sleeveless hostess gown. Labeled: Andrea. ($40-45)

Pink, gray, lilac, green, and white "splash of flowers" sleeveless hostess gown. ($40-45)

Hostess gown with a black, long-sleeved bodice and bright pink, green, and black floral print skirt. ($40-45)

Pink, green, and white floral print, long-sleeved hostess gown with a black velvet sash and buttons. ($35-40)

Green and white "geometric ivy" print, long-sleeved hostess gown. ($45-50)

Opposite page: Blue, white, yellow, and orange "geometric daisy" print, long-sleeved dress. ($65-85) *Modeled by Beth Geronikos.*

Black, red, orange, and white "trees and flowers" print, long-sleeved hostess gown. ($50-55)

Aqua, orange, and yellow "daisy" print, long-sleeved dress. Labeled: Honey Comb, "Bee Comes You." ($24-28)

FRUIT
ORANGE

Tan, lilac, blue, and rust "outer-space" print coat dress. ($75-85)

Aqua, blue, and pink print with a "ribbons and fans" motif; long dress with matching jacket. ($95-100)

Gray short sleeved, two-piece "commuters on the subway" dress. Labeled: Huk-a-Poo. ($65-75)

Blue, white, and brown "decorated clay pitchers" print skirt and long-sleeved, tie-front jacket. ($65-70 set)

Below:
Pink, mauve, and magenta "marble swirl" print, long-sleeved, zipper-front robe. ($35-40)

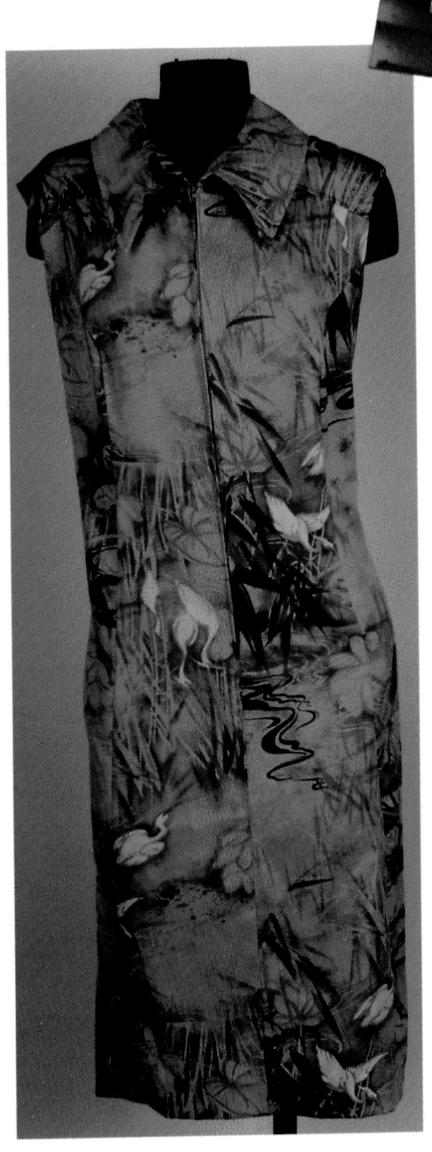

Lime green and white "flamingos in a pond" print sleeveless dress. Labeled: Lorac Original. ($35-40)

Animal Skin Prints

A brown and creme "zebra" print dress. Labeled: A Gloria Swanson Fashion by Puritan Forever Young. ($65-75)

A tan and brown "tiger-striped" gown. ($70-75)

Brown, tan, and white "leopard" print dress. ($65-70)

Solid black skirt with tiger print top in shades of black, tan, and white. ($60-65)

Black and white abstract print dress. ($75-85)

Pink, red, orange, purple, and white "dall sheep" print, long-sleeved dress. ($65-75)

Tan "leather-like" dress with sleeves and collar in a tiger print. Entirely of polyester, this mini-dress has the feel of soft leather. ($65-70)

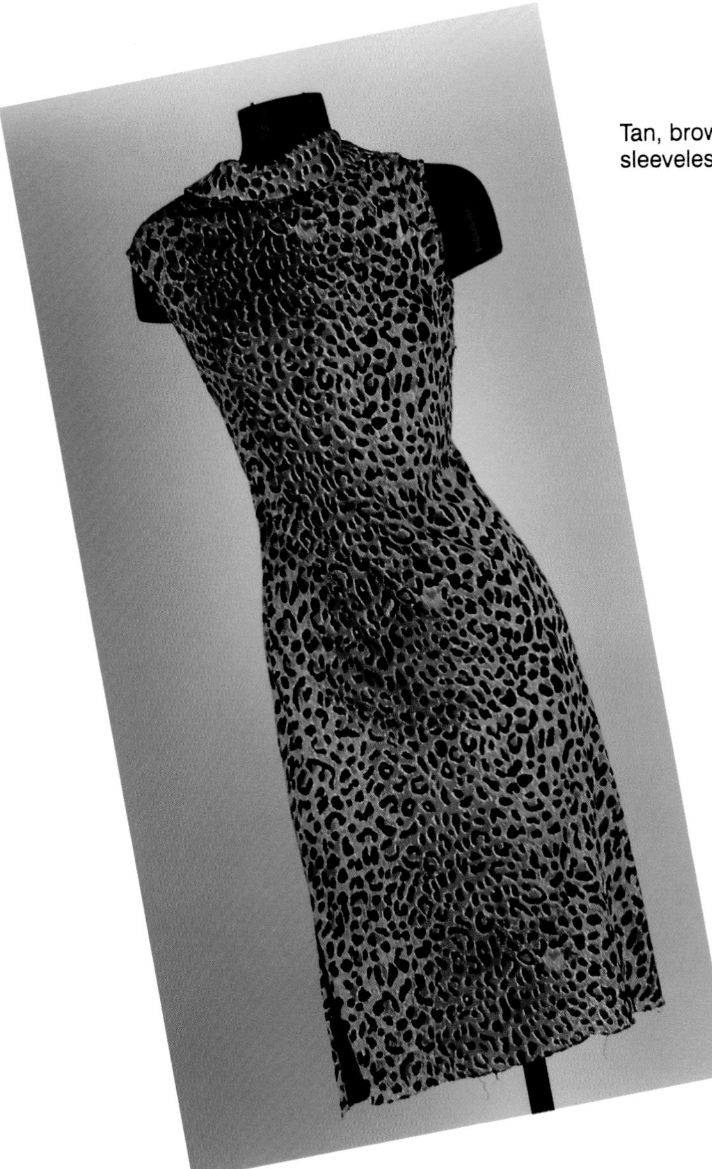

Tan, brown, and black leopard print sleeveless dress. ($35-40)

Brown and tan "leopard" print dress trimmed with black. ($40-45)

Stripes

Yellow gown with yellow, red, white, and blue striped halter top. ($65-70)

A white "tennis" dress with maroon and gray striped trim. ($40-45)

Blue, white, and yellow striped mini-dress with a pleated skirt and solid blue top, trimmed with striped polyester. Labeled: 100% polyester, a POLAR dress. ($45-50)

Peach and aqua striped mini-dress. ($25-30)

A yellow, brown, and white "zig-zag" print, zipper front mini-dress. ($30-35)

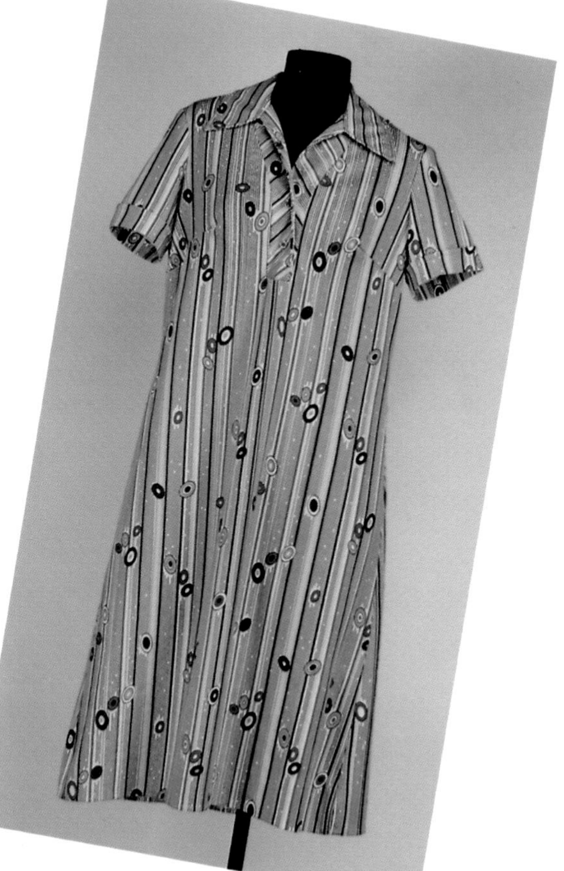

Blue, gold, white, and green "stripes and circles" dress. ($40-45)

Brown, blue, orange, and gold print strapless dress with an elaborate drape. ($65-75)

Maroon, blue, and white sheer sleeveless dress with matching jacket. ($50-55)

Navy blue, yellow, and white horizontal striped short-sleeved dress. Labeled: Laura Lenox. ($35-40)

Black, orange, brown, and creme "stripes and squares" dress. Labeled: Flair. ($50-55)

Red, white, and blue "circle" print long-sleeved dress. Labeled: Leslie Fay. ($40-45)

Sleeveless dress in wide stripes of navy blue, red, yellow, and white. ($30-35)

Red and white zig-zag print dress with red and white plastic "diamond" dangling from the front zipper. ($25-28)

Blue, orange, yellow, and aqua "v-shaped" stripes on a black ground. Labeled: Sears. ($35-40)

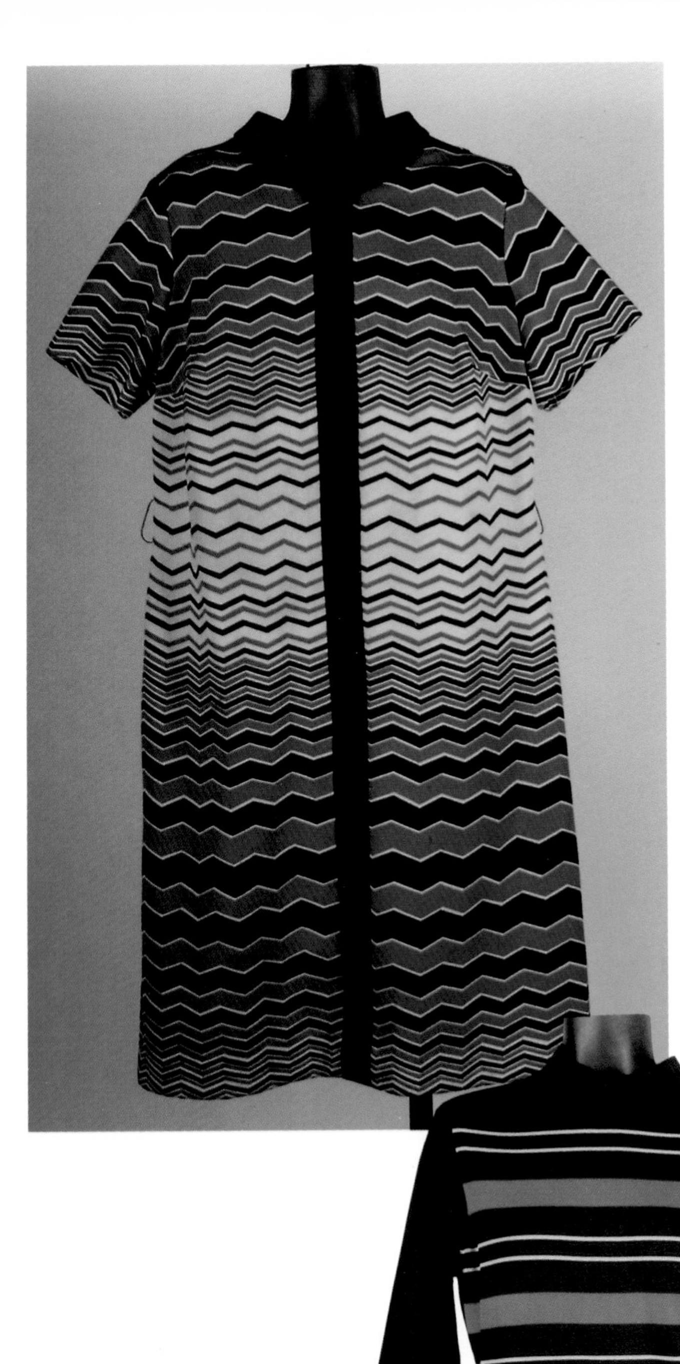

Green, white, and navy blue "zig-zag stripe" print short-sleeved dress. ($30-35)

Green and white striped robe, with an orange, pink, and brown "leopards lurking" print bodice and sleeves; additional leopard print down the wrap front. ($45-50)

Navy blue long-sleeved dress with fuschia and white horizontal stripes across the bodice. ($28-30)

Geometric Prints

Blue, black, and white geometric, rickrack print dress. Labeled: a Westover Walker dress. ($85-90)

An "easy-wear" zipper front dress in a red, blue, and black "circle" print on a white ground. ($55-60)

A long, geometric print dress in shades of red, yellow, green, and blue with a choker collar and slit front. Labeled: 100% polyester. ($65-70)

Blue and white geometric print dress with a circle motif and a solid blue ruffle at the hemline. ($55-60)

Pink, blue, and rose "marble-swirl" dress. Labeled: A Leslie Fay Original, 100% polyester. ($45-50)

A black dress with a floating "stained glass staircase" in shades of blue, black, pink, and orange. Labeled: a Lady Blair dress, 100% polyester. ($85-95)

A long geometric print dress in shades of orange, blue, green, gold, purple, and white. ($85-90)

Black, brown, and white geometric print, an intricate "pick-up-sticks" pattern. ($55-60)

Colorful Navajo print in shades of deep orange, yellow, blue, and green. Labeled: 100% polyester. ($70-75)

Purple, blue, and gray geometric print dress, a "design within a circle" motif. ($85-90)

Top and matching skirt in ice blue with "neon-colored" geometric shapes. ($95-120 set)

Sleeveless gown in shades of red, yellow, orange, and purple; a "honeycomb" blend. Labeled: A Belleza gown, 100% polyester. ($90-95)

Opposite page:
Top and matching skirt in ice blue with "neon-colored" geometric shapes. *Modeled by Beth Geronikos.* ($95-120 set)

Long white dress with circular "air-brushed" designs in shades of rose, pale green, and black. *Modeled by Katherine Wellington.* ($75-80)

Long dress with red, black, and white print skirt and solid black long-sleeved top with a scoop neckline. Labeled: Eyefull. ($55-60)

Long sleeveless dress with purple and white print top and solid purple skirt. ($65-75)

Black and white "rough seas" geometric print dress & sash. Labeled: Sears. ($45-50)

Black, white, and yellow geometric print long-sleeved dress. Labeled: Hob-nobber naturally. ($40-45)

Brown, white, and gold "window through the clouds" print long-sleeved dress. ($30-35)

Below:
Long-sleeved dress, with a red "vest style" bodice with attached white sleeves and collar, and a red and white plaid skirt. ($35-40)

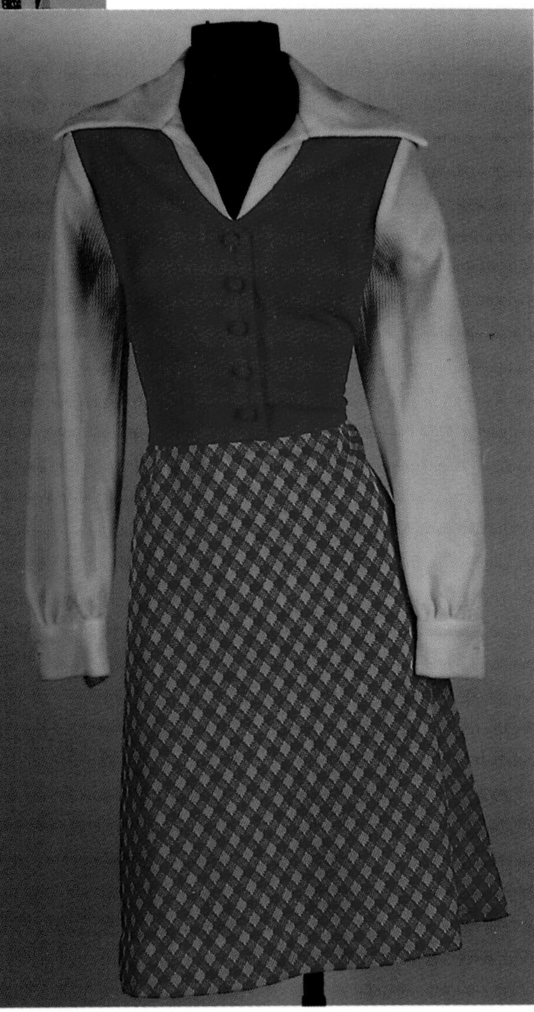

Left:
Crepe dress with solid "chocolate" top with creme and brown geometric "grid" skirt. Labeled: Fred Rothechild of California. ($55-65)

Long dress in shades of blue with an "artist's splash" of color. ($75-80)

Lilac "spaghetti straps" dress, with a permanently pleated skirt. *Modeled by Katherine Wellington.* ($40-45)

Opposite page:
Pale blue mini-dress. *Modeled by Katherine Wellington.* ($40-45)

Gold, long-sleeved knit dress with smocking on the shoulders and cuffs, with sash. ($55-65)

Mandarin orange Oriental gown with white detailing. ($55-60)

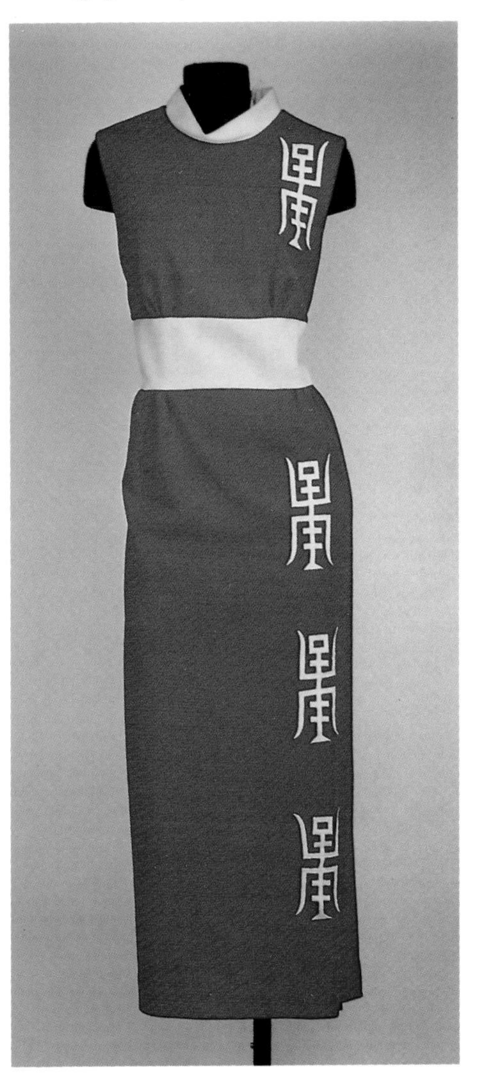

Opposite page:
Red/orange mini-dress with white buttons and stitching. *Modeled by Kayla Agran.* ($40- 45)

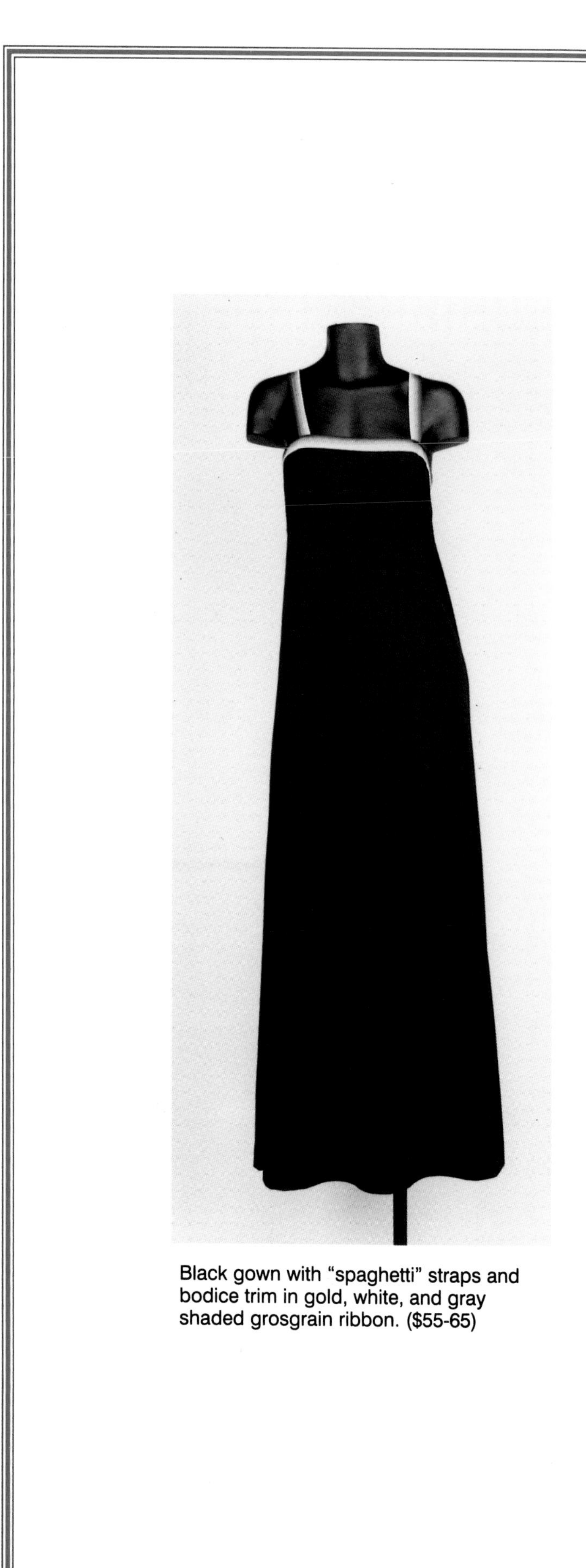

Black gown with "spaghetti" straps and bodice trim in gold, white, and gray shaded grosgrain ribbon. ($55-65)

Long purple sleeveless dress with open-work at neckline; side-slits and belt trimmed with pale blue and red "diamond" print. ($75-85)

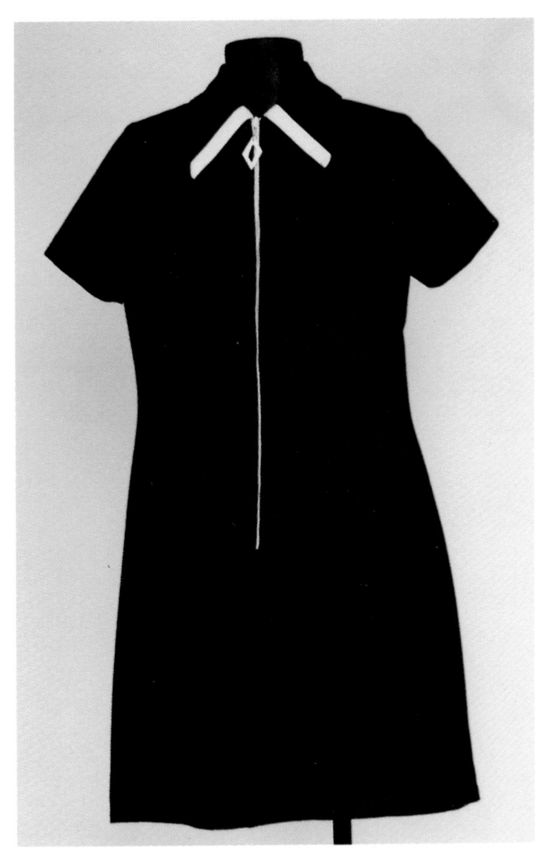

Navy blue mini-dress with a white zipper and collar trimmed with white. ($45-50)

Pink and white knit long-sleeved dress with open-work at the neckline. ($50-55)

Brown mini-dress with white piping, square white buttons, and a white patent leather belt. ($45-50)

Black dress with snowflake design collar and belt. ($35-40)

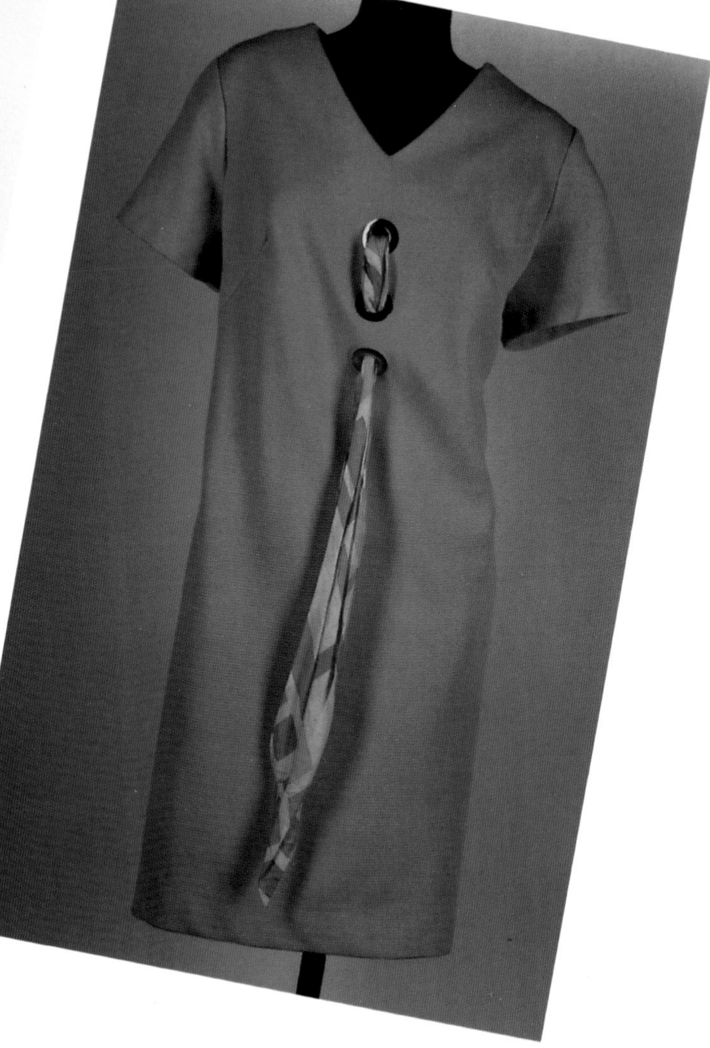

Lime green short-sleeved dress with mauve, pink, and white "belly button" sash. ($35-40)

Brown and white sundress with "spaghetti straps." ($35-40)

Navy blue and white sleeveless tennis dress. ($24-28)

Navy blue short-sleeved dress with red and white detailing at the neckline. Labeled: Leslie Pomer. ($24-28)

Black long-sleeved dress with red detailing and buttons; red and black sash. ($24-28)

Black and white halter-style gown. ($75-95)

Chapter Two
Shirts & Blouses

Sleeves

Most of the polyester men's shirts and women's blouses of the 1970s have tight-fitting sleeves with tailored, flat seams set smoothly into the armholes. Sometimes the sleeve has two small, unpressed pleats where it joins the cuff and many of these shirts have one or two buttons on the cuff. Sometimes the shirts have bell-sleeves which are narrow at the top and flare at the lower edge like a bell. Women's blouses are cut similar to men's shirts, but some are sleeveless. Other blouses have very small cap sleeves in which an extension is cut on the front or back of the blouse to cover the shoulder.

Styling

The vast majority of polyester shirts and blouses button up the front. However, pullover styles, with long or short sleeves, are also found. The designs are so diverse and fascinating, that to present over one hundred carefully chosen examples is just the tip of the iceberg!

Fabric Designs

The most important feature of a polyester shirt or blouse is the fabric design. Enjoy the diversity you find in these representative examples.

Floral Prints

Ivory sheer long-sleeved shirt with a "sea of faces" in blue. Labeled: A Jones blouse, 100% polyester. ($28-32)

Brown, tan, yellow, and white floral, long-sleeved shirt. Labeled: Cortini Knit Shirt. ($22-28)

Blue on blue floral print, long-sleeved shirt. ($12-15)

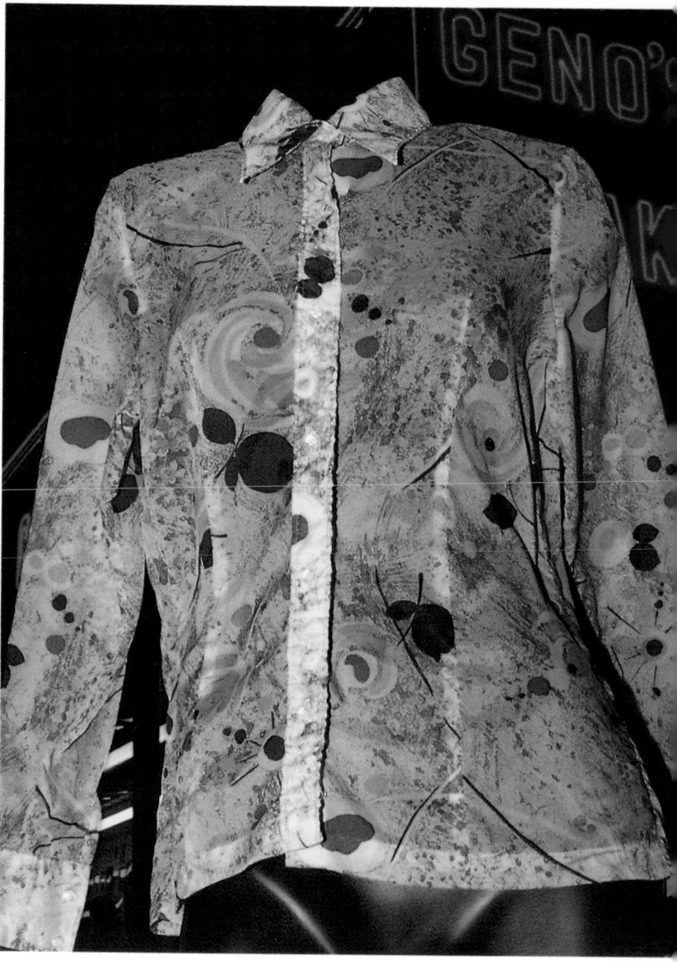

Sheer, peach-colored, long-sleeved shirt with a floral pattern in shades of blue. Labeled: 100% polyester. ($24-28)

Floral print, short-sleeved shirt in shades of green. Labeled: Haband. ($18-22)

Floral print, long-sleeved shirt in shades of blue. ($18-20)

Blue, long-sleeved shirt with "roses on a trellis" motif. ($28-32)

White, green, and red "herbs and flowers," long-sleeved shirt. Labeled: Fred Dunni. ($22-24)

Yellow, orange, purple, and white floral print, long-sleeved shirt. Labeled: Stage 7. ($18-20)

"Asian house of flowers," short-sleeved shirt in shades of white, orange, black, pink and green. ($24-28)

"Flower garden," short-sleeved shirt in shades of white, orange, yellow, and green. ($22-24)

"Wild vines and watermelons," long-sleeved shirt in shades of red, white, green, and mauve. Labeled: Motivation ($30-35)

Long-sleeved shirt with red flowers and blue diagonal stripes. Labeled: Donegal. ($18-22)

Shades of pink "Greek Goddess in the garden," long-sleeved shirt. Labeled: Loubella extendables. ($25-28)

Creme, orange, brown, blue, and black "mysterious woman," long-sleeved shirt. Labeled: A Sybil shirt. ($20-24)

Blue and coral "boating," short-sleeved shirt. Labeled: 100% polyester. ($25-28)

Peach colored "bon voyage," long-sleeved shirt. ($18-20)

Long-sleeved shirt with red and white flowers and geometric shapes on a peach ground. ($24-28)

Orange, yellow, brown, and white, long-sleeved shirt with "bricks and flowers" design. Labeled: Shapely. ($24-28)

Blue, gray, and white "faux vest," long-sleeved shirt. ($28-32)

Black, tan, and ivory floral print, short-sleeved Hawaiian shirt. Labeled: Otaheite from Maui. ($32-35)

Tan, green, and creme "peacock feather" print, long-sleeved shirt. Labeled: Zoology. ($28-32)

Black, white, and red "honeycomb with poppies" print, long-sleeved shirt. Labeled: Leonard Strassi Move with the Times. ($32-35)

Peach, brown, and creme "abstract floral" print, long-sleeved shirt. ($24-28)

Pink, blue, green, and tan sheer "forest delight" print, long-sleeved shirt. Labeled: h.i.s. for her. ($28-30)

Navy blue and white floral print, long-sleeved shirt with two patch pockets. Labeled: Vera. ($24-28)

Beige, gray, and tan "birds and flowers" print, long-sleeved shirt. ($22-24)

White, blue, and tan floral print, long-sleeved shirt. ($20-22)

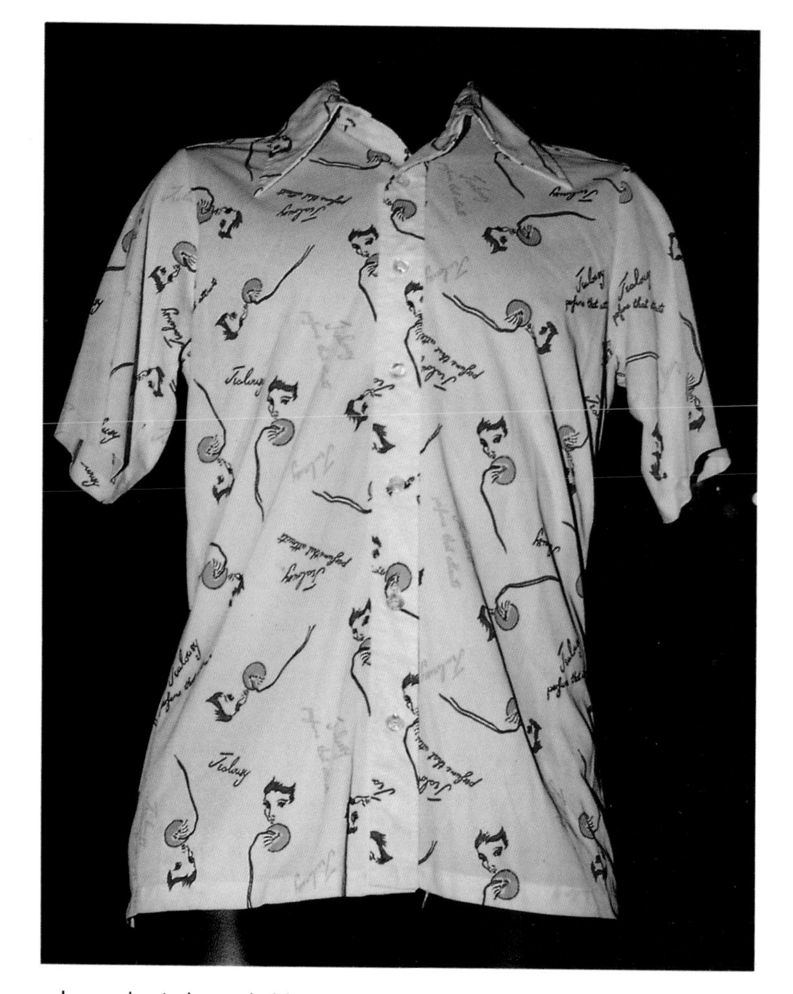

Ivory short-sleeved shirt with red, black, and yellow design and printing, "Jealousy. Perfume that attracts." Labeled: Arnold's Men's Shop. ($32-35)

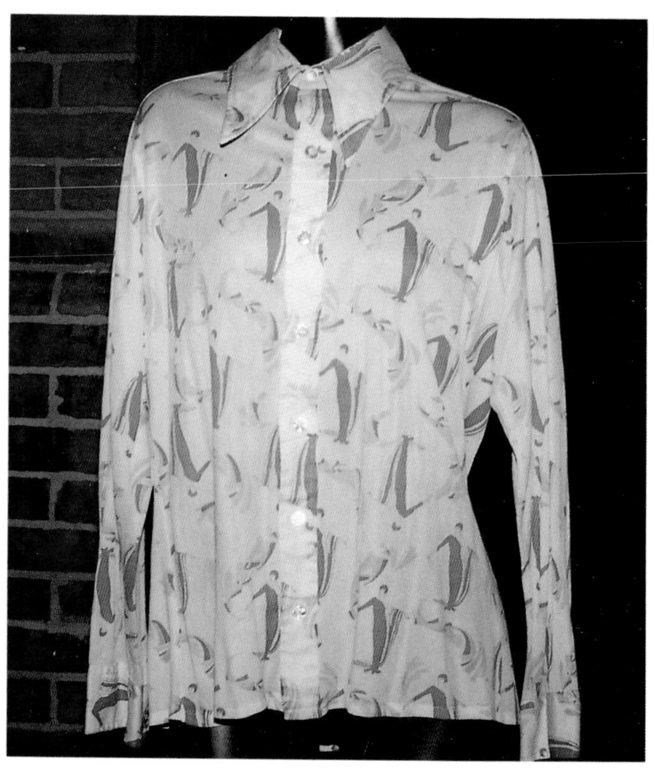

A long-sleeved shirt with "Art Deco" print of a woman in shades of orange on an ivory ground. Labeled: 100% polyester. ($32-35)

Sage green, long-sleeved shirt with "forest scene" in shades of pink, green, and brown. Labeled: It's Gailord. ($18-20)

Blue and coral "boating," short-sleeved shirt. Labeled: 100% polyester. ($25-28)

Peach colored "bon voyage" long-sleeved shirt. ($18-20)

Orange, creme, and tan "deserted town," long-sleeved pullover. ($25-28)

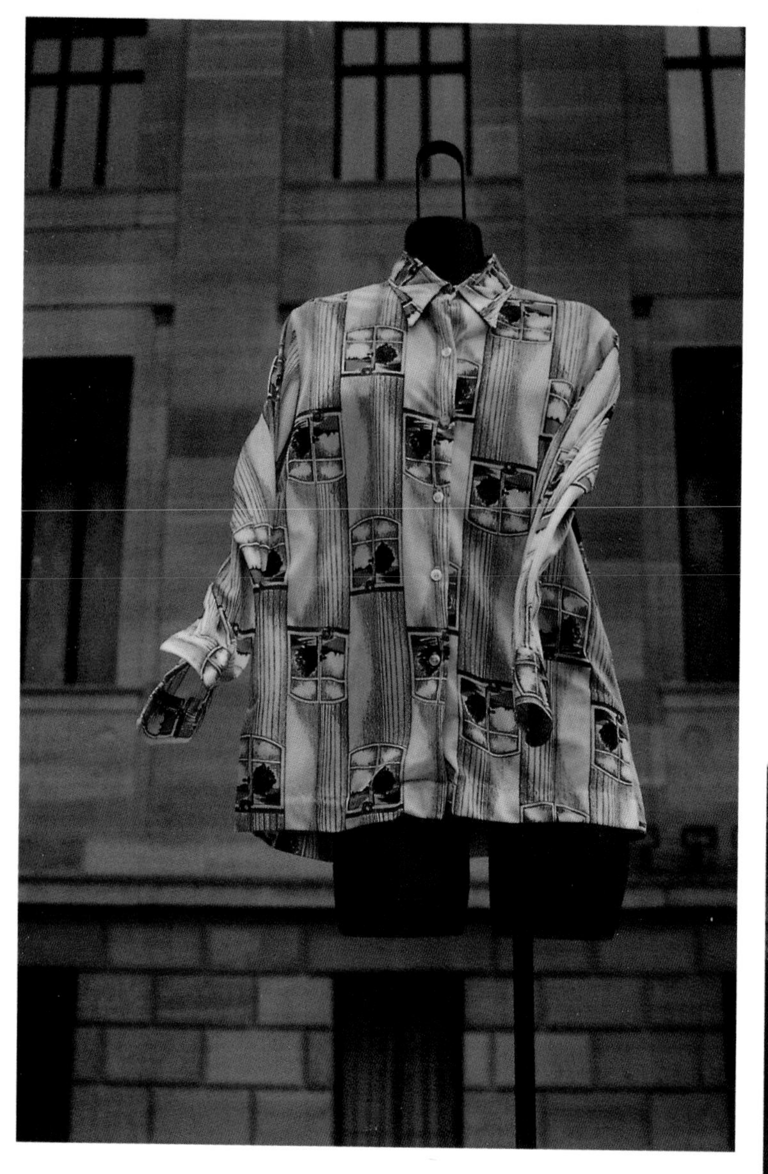

Tan, brown, and creme "a room with a view," long-sleeved shirt. Labeled: Gabey. ($28-32)

Short-sleeved shirt in shades of brown, tan, and orange; "vertical lines," Labeled: John Wanamaker. ($24-28)

Blue, brown, and orange "images of women," long-sleeved shirt. Labeled: New Quinessa Fabric. ($28-32)

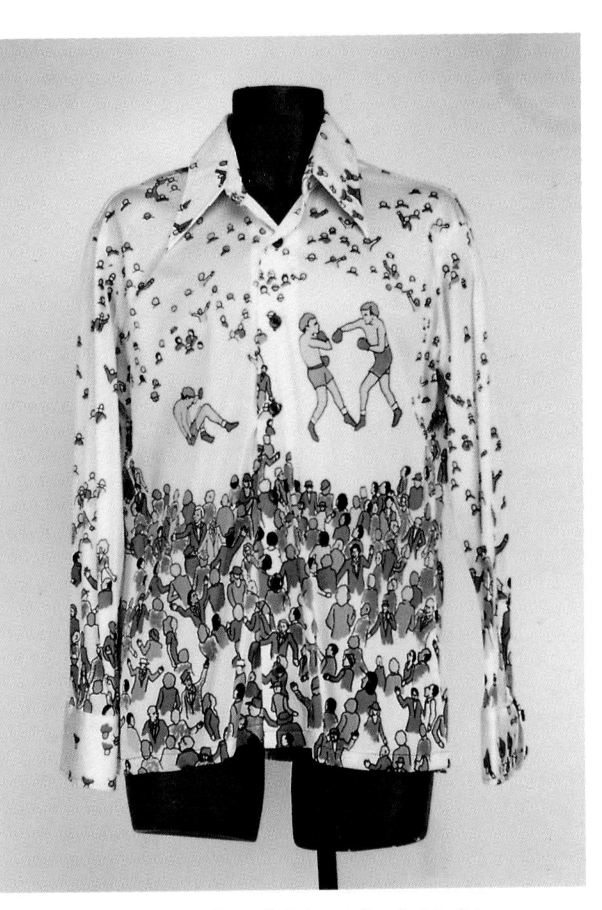

Creme, blue, and tan "night at the fights," long-sleeved shirt. Labeled: h.i.s. ($32-35)

Aqua and white "fat daddy caddy shack," short-sleeved pullover. Labeled: Robert Bruce Arnold Palmer Design. ($28-32)

Rust and creme "Roman figural" print, long-sleeved shirt. Labeled: Devon. ($24-28)

Creme, blue, maroon, and black "newspaper" print, short-sleeved shirt, front-tie. Labeled: Laffin Lass. ($30-32)

Colorful "happy cafe," short-sleeved pullover. Labeled: Ship N Shore. ($24-28)

Creme, tan, blue, and orange "riding across the plains" print, long-sleeved shirt. Labeled: Personal. ($28-32)

Gray/blue "ocean view" print, long-sleeved shirt. Labeled: Accent. ($28-30)

Brown, blue, and white "architectural" print, long-sleeved shirt. Labeled: h.i.s. ($28-30)

Gray, orange, and green "volcano" print, long-sleeved shirt. Labeled: Eastbay. ($35-40)

White, green, and black "horse and rider" print, long-sleeved shirt. Labeled: Catalini. ($30-35)

White, green, orange, and pale yellow "abstract ocean waves at sunset" print, sleeveless shirt. ($20-24)

Navy blue long-sleeved shirt with "sailboats, anchors, and dots" print collar and cuffs. ($20-24)

Dark green, tan, and white "deep forest" print, long-sleeved shirt. Labeled: Peception directions. ($35-40)

Tan, gray, blue, and white "women picking roses" print, long-sleeved shirt. Labeled: Huk-a-Poo. ($30-35)

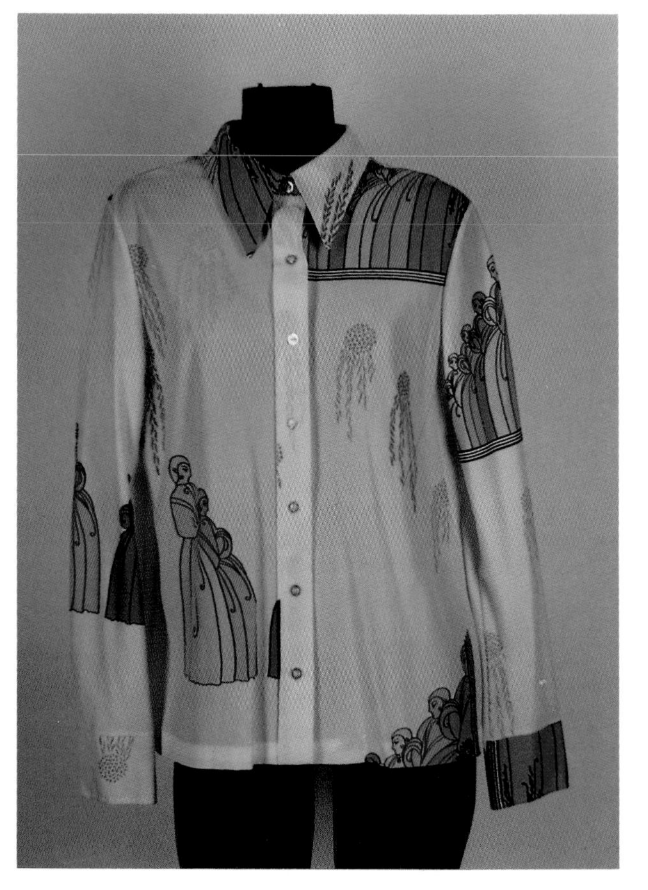

White, orange, pink, and aqua "colonial woman" print, long-sleeved shirt. Labeled: Devon. ($30-35)

Aqua, baby blue, lilac, and white "dream sequence" print, long-sleeved shirt. ($30-32)

Red, white, and pale blue horizontally striped, short-sleeved shirt. ($28-30)

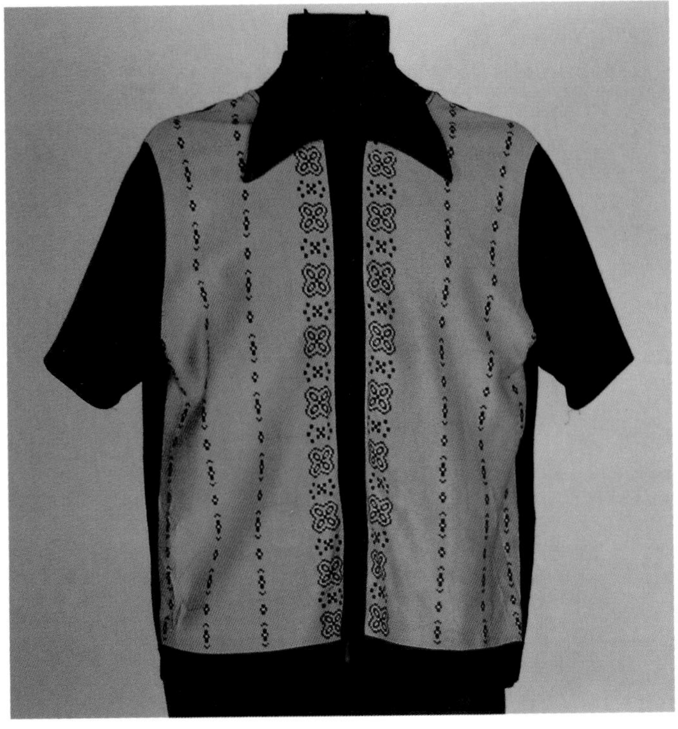

Brown and beige "mock vest" style, short-sleeved, zipper-front shirt. Labeled: Enrico felini. ($45-50)

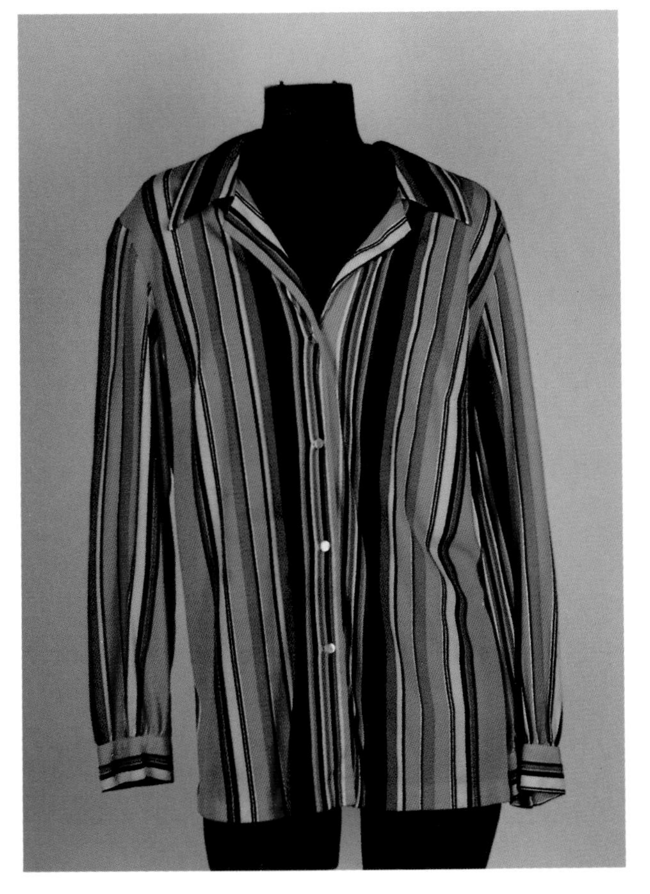

Yellow, blue, red, and white vertically striped, long-sleeved shirt. Labeled: Stockton. ($22-24)

Navy blue, red, and white "zebra" striped, long-sleeved shirt. ($20-22)

Red, blue, tan, and white "graduated paisley" print, long-sleeved shirt. Labeled: Hooper. ($28-30)

Black long-sleeved shirt with metallic gold, horizontal stripes; drawstring waist. ($30-35)

Brown and white vertical stripes, long-sleeved shirt. Labeled: Act III. ($18-20)

Green, orange, and creme Oriental "fan" motif, long-sleeved shirt. ($18-20)

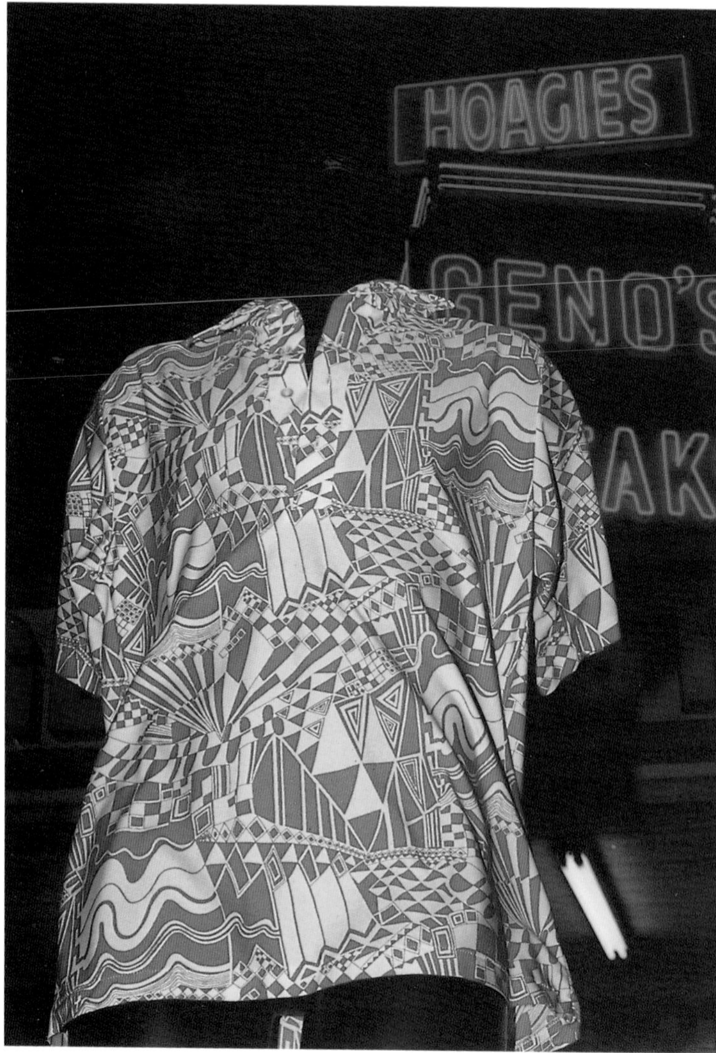

Red and white men's short-sleeved shirt with geometric pattern and "checkers." Labeled: 100% polyester. ($22-24)

Red, blue, yellow, green, and white geometric print long-sleeved shirt showing the influence of Cubism. Labeled: 100% polyester. ($32-35)

Brown, tan, and orange long-sleeved shirt with geometric "zig-zag " design. Labeled: A Kay Windsor blouse. ($18-20)

Turtleneck, long-sleeved shirt with orange, brown, blue, and white "marble" design. Labeled: 100% polyester. ($30-32)

Pullover, long-sleeved shirt with brown diagonal stripes and green, yellow, orange, and brown circles; back zipper. ($32-35)

Pale green, long-sleeved shirt with a "grid" of darker green and a design of geese flying high above tree branches. Labeled: 100% polyester, Trissi's women's blouse. ($18-20)

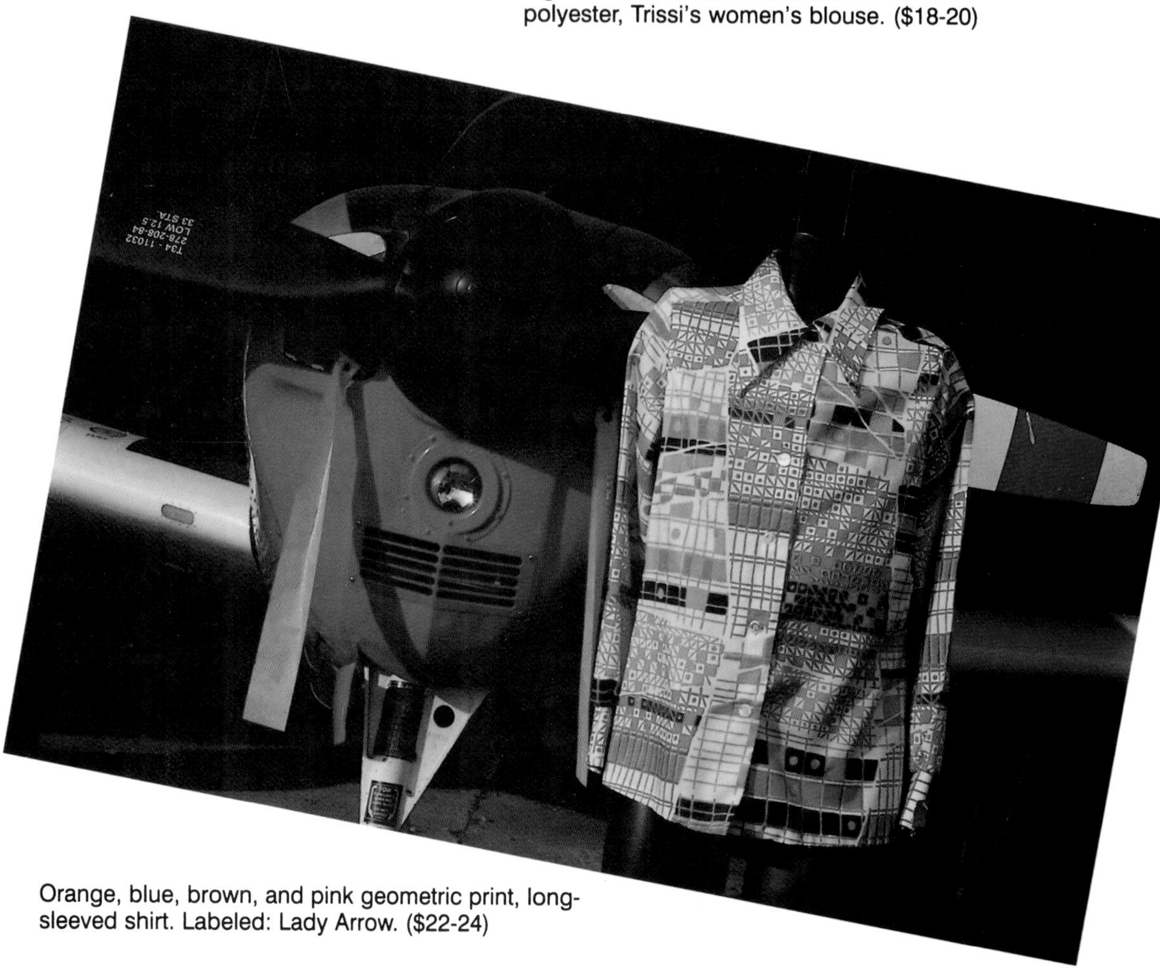

Orange, blue, brown, and pink geometric print, long-sleeved shirt. Labeled: Lady Arrow. ($22-24)

Floral print, long-sleeved shirt in shades of rust and blue. Labeled: Romelli. ($18-20)

Ivory, short-sleeved shirt with a geometric design in shades of brown, rust, and yellow. Labeled: Arrow. ($22-24)

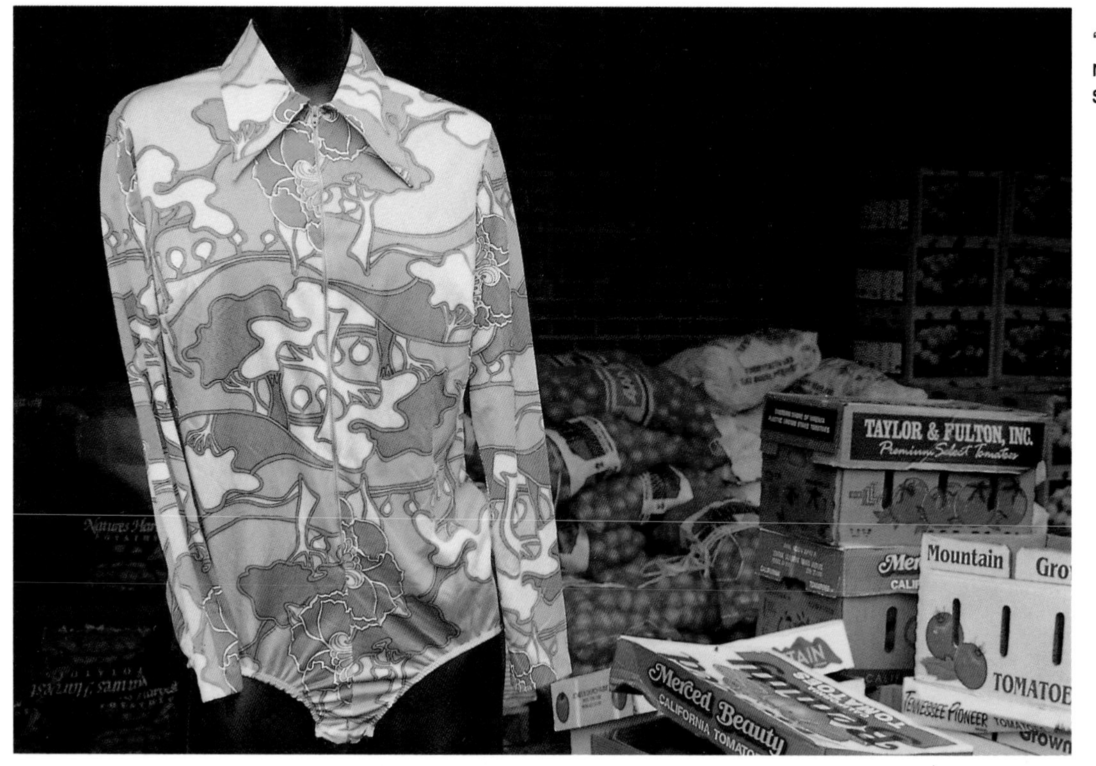

"Follow the yellow brick road," long-sleeved body suit. ($35-40)

Black and white "Oriental circle," bell-sleeved shirt. Labeled: Trissi. ($24-28)

Neon green, pink, and yellow "crazy square" short-sleeved shirt. ($22-24)

Hot pink and green, sleeveless shirt with white lines and circles. ($15-18)

"South Florida development plan" design, long-sleeved shirt. ($28-32)

Aqua, black, and white plaid, tie-front, long-sleeved shirt. ($15-18)

Brown and white polka dot, long-sleeved shirt. Labeled: A King's Road shirt. ($22-24)

Red, white, and blue "daisy within a circle" design, long-sleeved shirt. ($32-35)

"Expressionistic" pink and blue floral print, long-sleeved shirt. ($28-32)

Opposite page:
Creme, black, maroon, and mauve "safety-pin" design, short-sleeved shirt shown with maroon pants. *Modeled by Peter N. Schiffer.* ($32-35)

Brown, creme, and orange "backyard boogie" print, long-sleeved shirt. Labeled: Leonardo Strassi. ($32-35)

Tan and creme "bridge on the horizon" print, long-sleeved shirt. Labeled: Roland. ($24-28)

Black, tan, gray, and brown "earth, wind, and fire," long-sleeved shirt. Labeled: D'Auica. ($35-40)

Orange, yellow, and brown, long-sleeved, floral patchwork shirt with ruffles at the cuffs. ($18-22)

Pale yellow with green "helmets" print, short-sleeved pullover. Labeled: Robert Bruce Arnold Palmer Design. ($28-32)

Peach, aqua, brown, and creme "waterfall of flowers" print, long-sleeved shirt. Labeled: Russ. ($28-30)

Pink, purple, lilac, green, and white "blowing bubbles" print, short-sleeved shirt. Labeled: Alfred Dunner. ($32-35)

White with red polka dots, short-sleeved shirt. Labeled: Givenchy of Paris. ($28-32)

Brown, rust, and creme bold geometric and floral print, long-sleeved pullover. ($18-20)

Black, creme, blue, and brown "abstract woman" print, long-sleeved shirt. Labeled: Sybil. ($28-30)

Red, tan, white, and blue "whirlpool" print, long-sleeved pullover. Labeled: Terry. ($28-32)

Pale blue with gold, green, and brown "Byzantine flair" print, long-sleeved shirt. ($24-28)

Red, blue, and gold "tapestry," long-sleeved pullover with a waist tie. ($28-30)

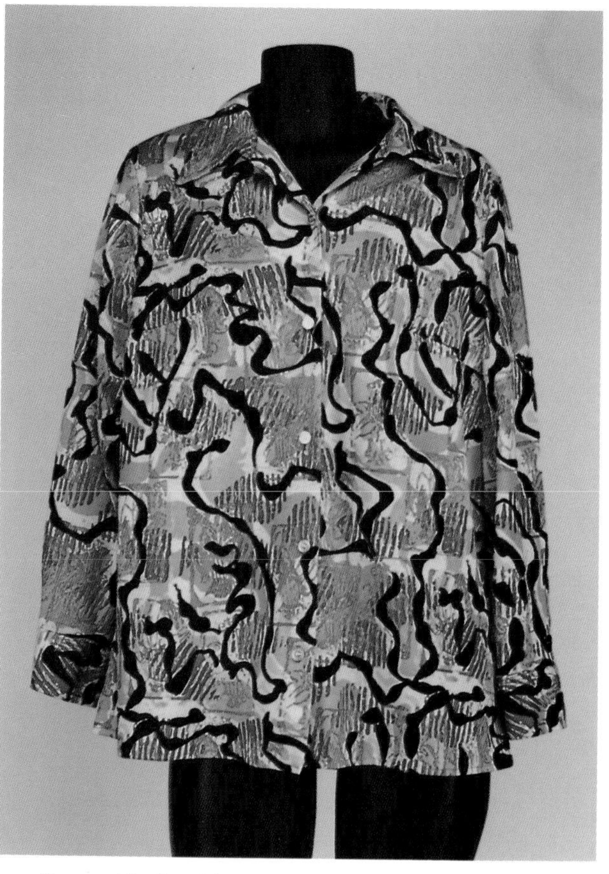

Green, black, white, and orange "abstract swiggle" print, long-sleeved shirt. ($28-30)

Pink, gray, and peach "geometric star" print, long-sleeved shirt. Labeled: Kokobay. ($28-30)

Brown, tan, and rust "vertically structured" print, long-sleeved shirt. ($20-24)

Creme, orange, black, and green "geometric floral" print, sleeveless pullover with four covered buttons at neckline. ($18-20)

Burgundy and tan polka dot print, long-sleeved shirt. Labeled: Devon. ($18-20)

Red and white polka dots, long-sleeved pullover with a zipper front. Labeled: Aladdin. ($15-18)

Long-sleeved vest-shirt with zipper front; solid maroon vest with maroon and beige "circle" print sleeves and collar.

Black, tan, and white "cubist" print, short-sleeved shirt. Labeled: LeChevron. ($28-30)

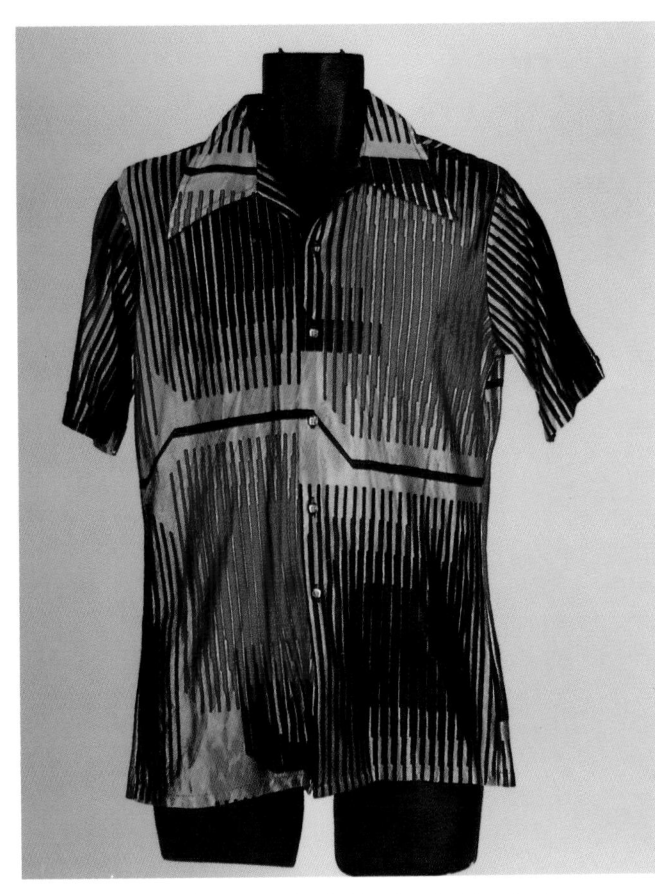

Black, red, and gray vertically "broken" striped, short-sleeved shirt. ($28-30)

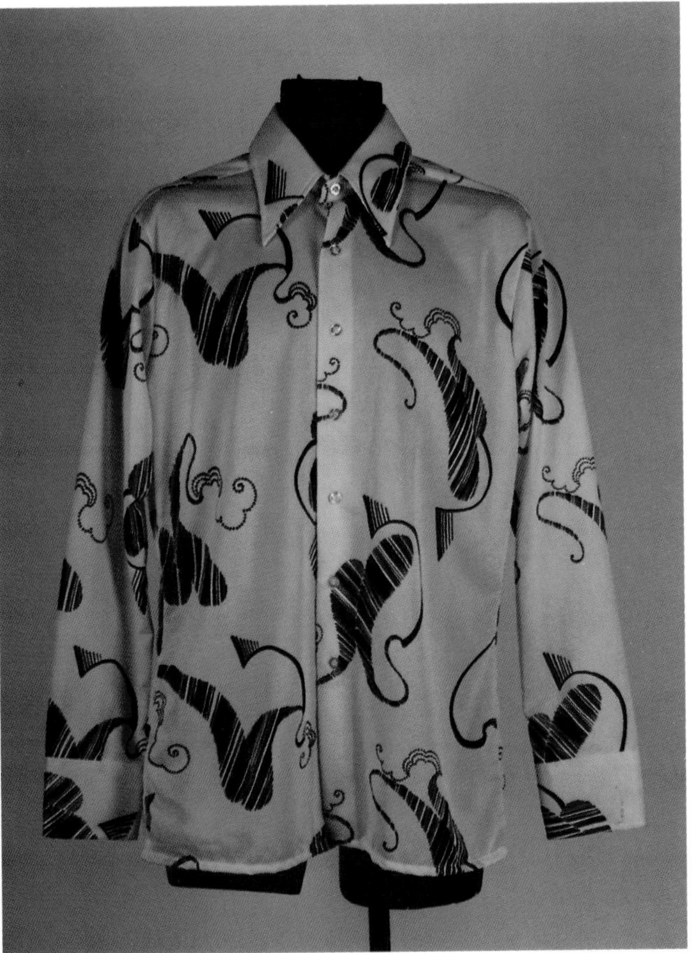

Black and white "abstract" print ,long-sleeved shirt. Labeled: Lucien Piccard. ($30-35)

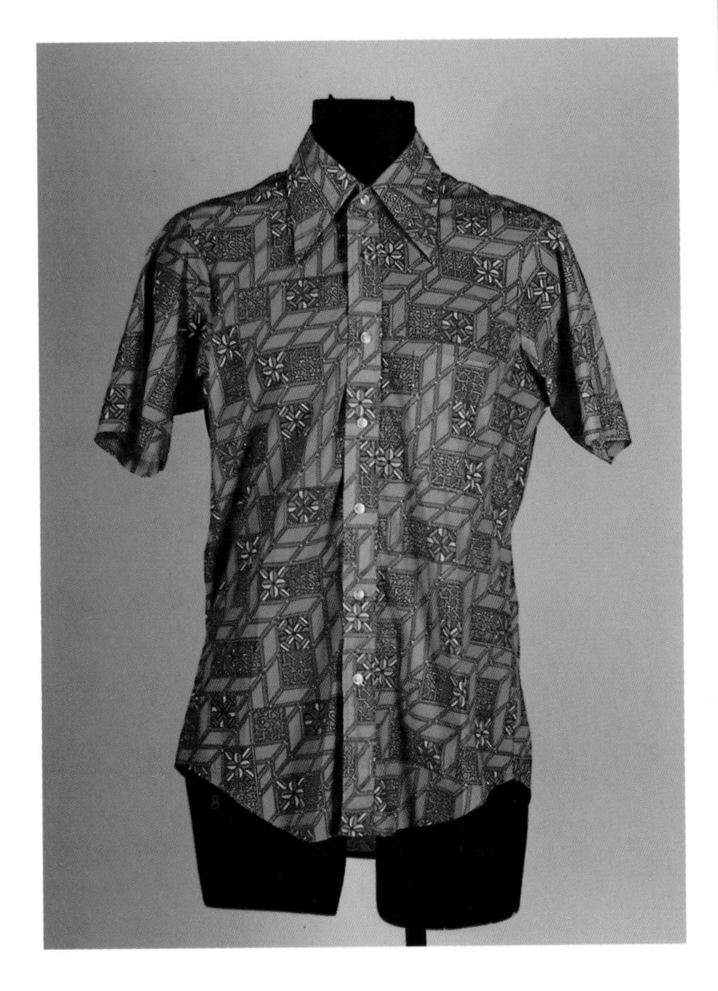

Mauve, blue, and red "stained-glass window" print, short-sleeved shirt. ($24-28)

Blue, yellow, and white "kaleidoscope" print, short-sleeved shirt. Labeled: Bravo. ($30-35)

Black and blue "tennis racket and net," long-sleeved shirt. Labeled: trissi. ($35-40)

Red, purple, and pale blue "diamond" print, long-sleeved shirt. Labeled: Gordon. ($22-24)

White, brown, green, and orange "kaleidoscope" design, short-sleeved shirt. ($30-35)

Two shades of green and white "fish scales" print, long-sleeved shirt. ($30-35)

Black, blue, red, pink, and white "squares within squares" print, long-sleeved shirt. ($35-40)

Black, tan, and blue "butterflies and three-dimensional figures" print, long-sleeved and matching short-sleeved shirts set. ($40-45 set)

Black, blue, pink, and white "moons in orbit" print, long-sleeved shirt. ($35-40)

Tan, orange, and yellow "pick up sticks" print, long-sleeved shirt. ($35-40)

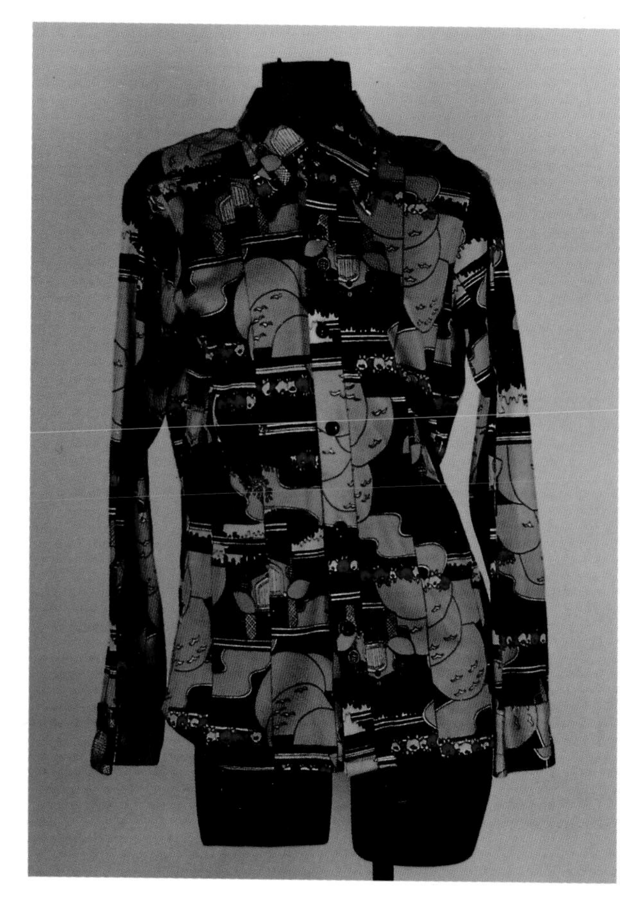

Navy blue, tan, red, and white "Model T" print, long-sleeved shirt. Labeled: Wrangler. ($30-35)

Black, pale blue, peach, and white "clouds in a dark sky" print, long-sleeved shirt. Labeled: Koko bay. ($28-30)

Black, gray, and blue "flowers in a field" print, long-sleeved shirt. Labeled: Roni Ren. ($28-30)

Blue pants suit with red floral print collar and cuffs on the short-sleeved jacket; same fabric sash. ($60-65 set)

Right: Floral print, two-piece swimsuit in shades of aqua, white, pink, and green. ($40-45)

Daisy print, two-piece swimsuit in shades of black, white, and yellow. ($50-55)

Blue and gray pin-striped, single-breasted suit. *Modeled by the author.* ($95-100)

Gold, long-sleeved shirt paired with gold, white, orange, and brown "zig-zag " print, flared-leg pants. ($65-75 set) *Modeled by Beth Geronikos.*

Three-piece layered look: tan turtleneck; brown button-down collar pullover; and blue, brown, and tan vertically striped jacket with horizontally striped patch pockets. ($10-12 turtleneck, $15-18 pullover, $65-75 jacket)

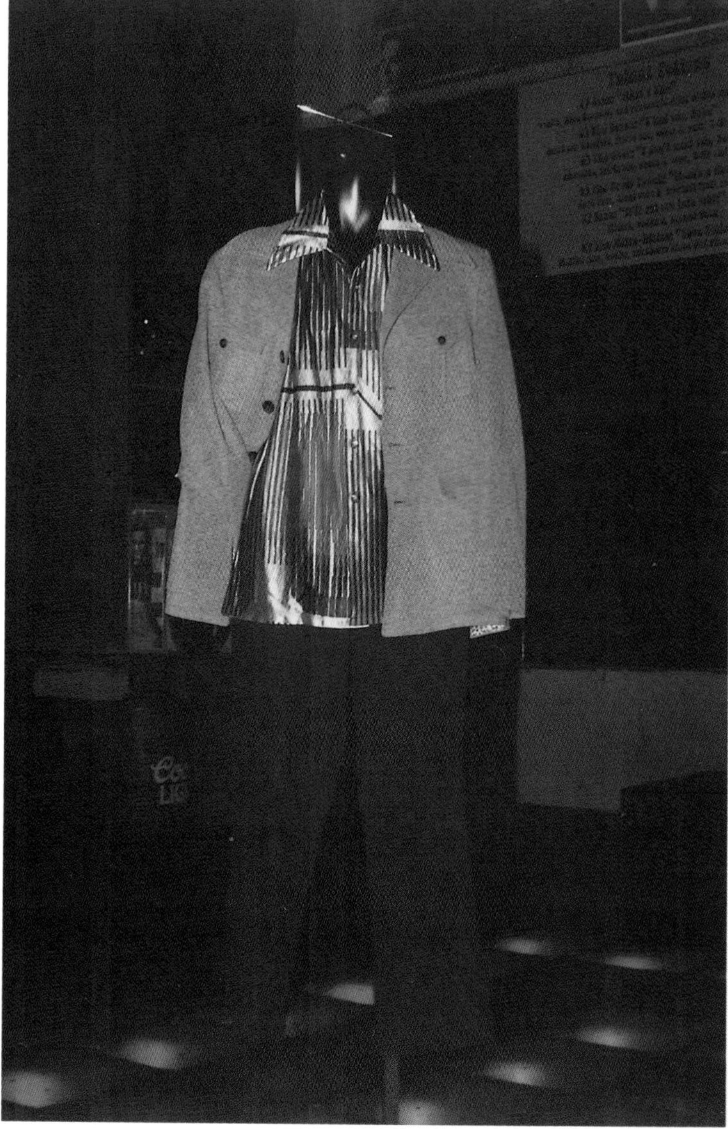

Blue double-knit jacket shown with red, white, and blue print shirt and solid navy blue double-knit pants. ($45-50 jacket, $24-28 shirt, $18-22 pants)

Red and black, pin-striped jacket. Labeled: Frank Colavito. Black, white, gray, and red "print of a woman" shirt. Labeled: Arrow Sportswear, shown with black pants. ($45-50 jacket; $24-28 shirt; $22-24 pants)

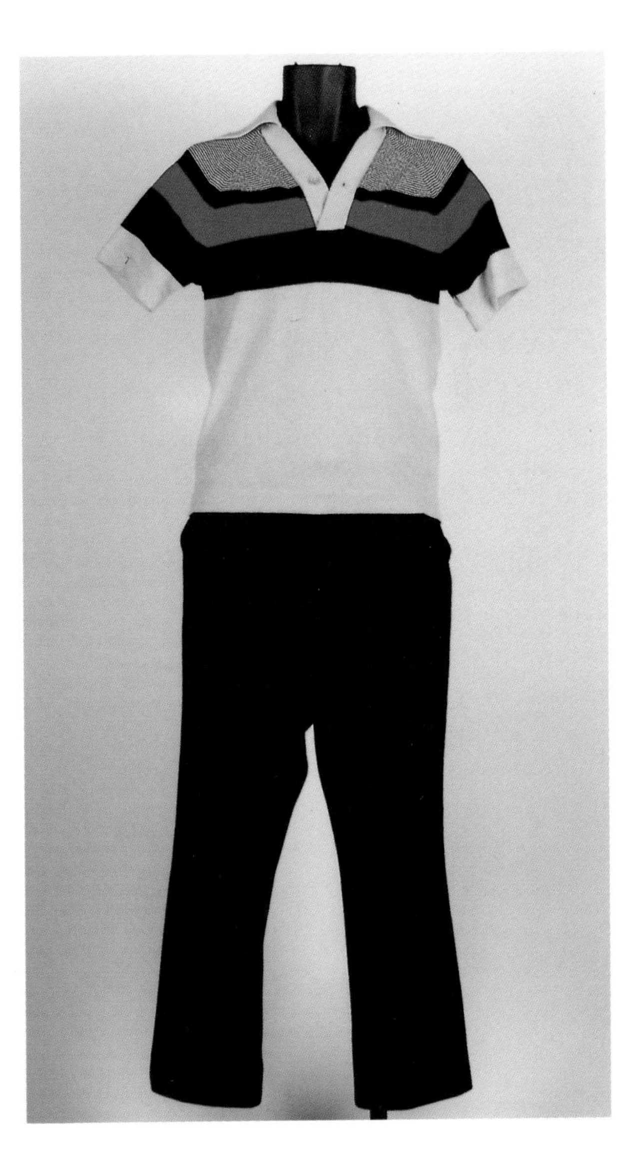

White short-sleeved shirt printed with bands of orange, black, and red across the shoulders, shown with black pants. ($20-22 shirt; $22-24 pants)

Geometric Prints

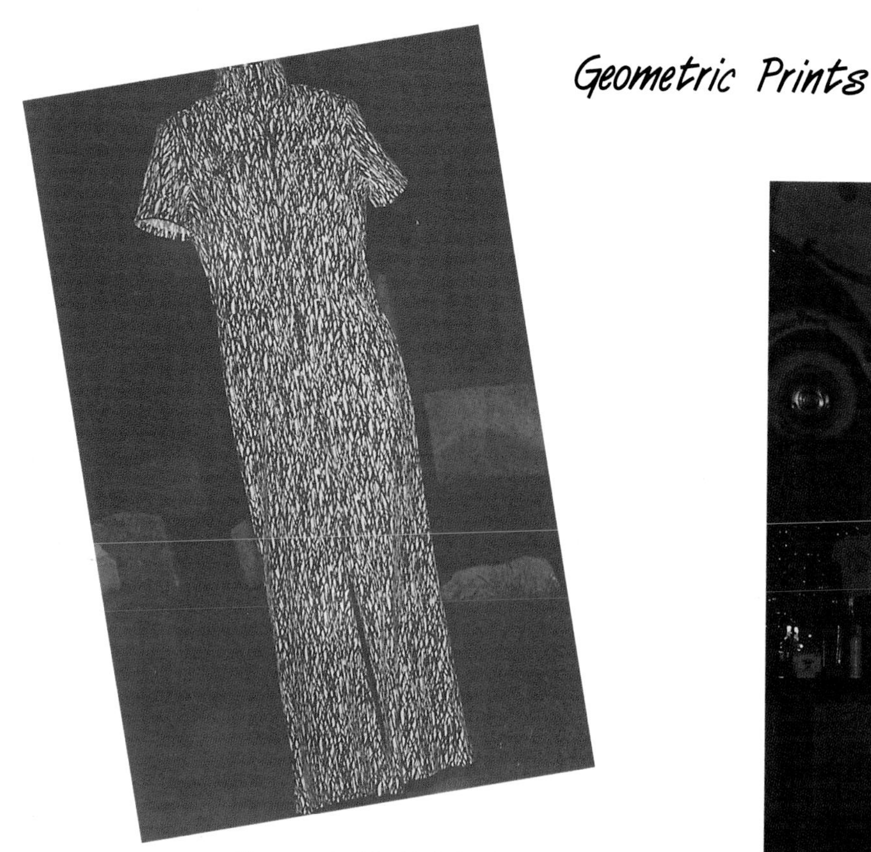

Black and white "twig" print, short-sleeved jumpsuit. Labeled: 100% polyester. ($55-65)

"Circle madness" pants suit in shades of black, red, yellow, and white; long-sleeved shirt with bell-bottom pants. ($95-120)

Brown, blue, yellow, and orange paisley print, sleeveless jumpsuit with harem pants. ($65-75)

Opposite page:
Solid green, flared-leg cuffed pants shown with a geometric print, long-sleeved shirt in shades of red, yellow, green, white, and blue; worn with high-platform, leather, multi-colored sandals. *Modeled by Katherine Wellington.* ($24-28 shirt; $28-32 pants)

Red and black print sports jacket trimmed with black velvet; shown with a solid red, button-down collar shirt and black pants. ($95-110 jacket, $15-18 shirt, $28-30 pants)

Red, white, and blue "houndstooth" knit pants suit with patch pockets and belt. ($40-45 set)

Opposite page:
Gray, blue, yellow, and white plaid pants ($32-35) and blue and silver tuxedo jacket. ($75-85) *Modeled by Peter N. Schiffer.*

Green double knit, white dots jacket with white piping at the pockets and two white buttons, shown with a white double-knit shirt. Jacket labeled: Jantzen. ($45-50 jacket, $15-18 shirt)

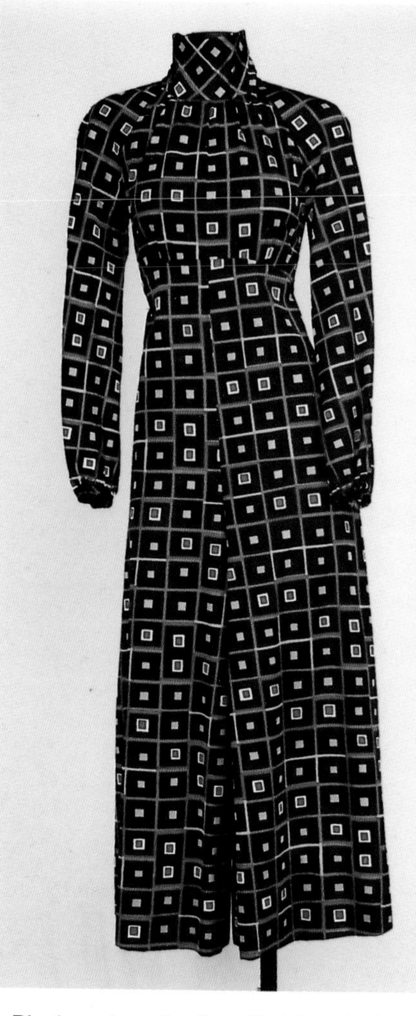

Black, red, and yellow "Rubix cube" print, long-sleeved jumpsuit. ($85-95)

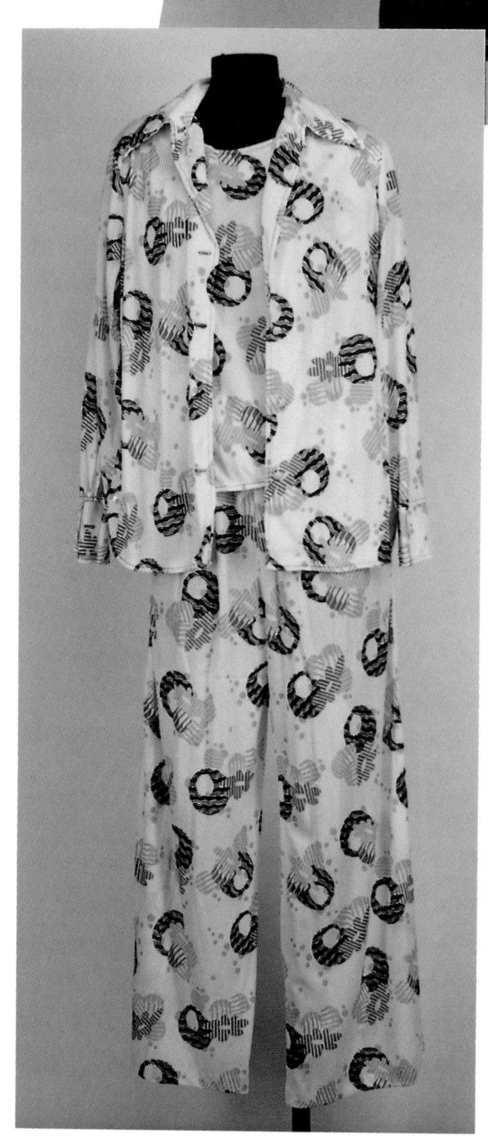

Abstract floral print ,three-piece pants outfit in shades of brown, mauve, and yellow on a white ground. ($65-75 set)

Opposite page:
Purple, white, black, blue, and yellow "geometric" print, halter-style jumpsuit. ($90-95) *Modeled by Beth Geronikos.*

PETET
1990six

Blue, white, and brown "pottery jugs" print, long-sleeved, flared-leg pants suit. ($90-95 set) *Modeled by Beth Geronikos.*

Brown, tan, and white floral print, long-sleeved jumpsuit. Labeled: Aviso. ($65-70)

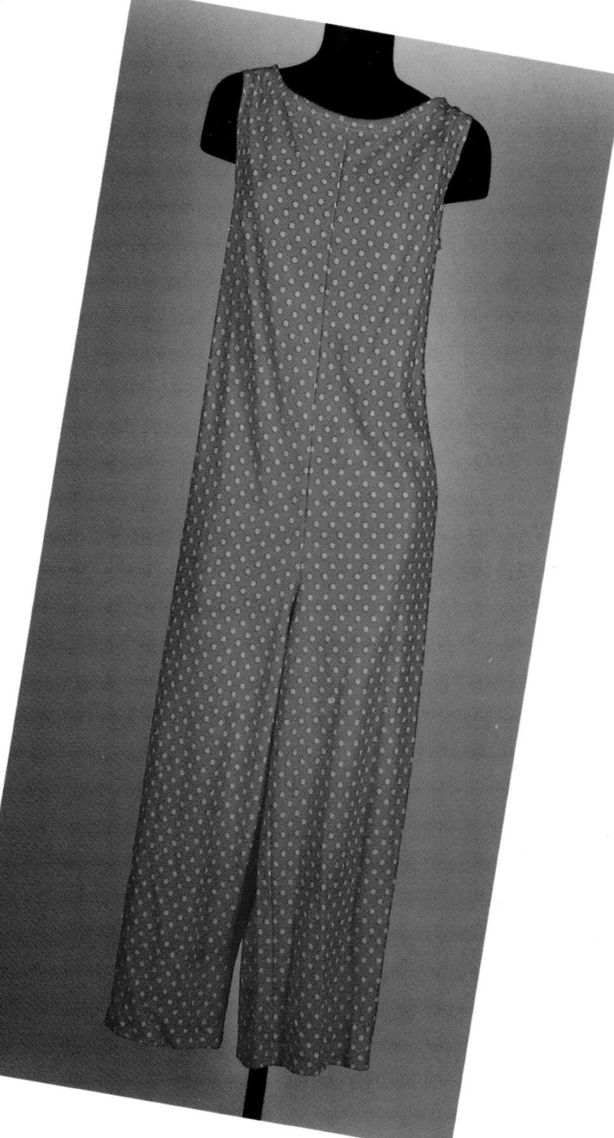

Pink, sleeveless jumpsuit with white polka dots. ($65-75)

Gray, light blue, and yellow plaid jacket over a yellow and gray "bird" print shirt; shown with black pants. ($65-70 jacket; $20-25 shirt; $20-22 pants)

Navy blue jacket with red-stitched patch-pockets. Labeled: Murrays "fit for kings." Red, white, and blue striped shirt shown with red, blue, and gold "patchwork" print pants. ($65-70 jacket; $15-18 shirt; $55-60 pants)

White jacket with green-stitched patch-pockets. Labeled: McGregor. White, green, and light blue "geometric" print shirt shown with green and white plaid pants. ($65-70 jacket; $22-24 shirt; $40-45 pants)

Blue jacket with darker blue stitching and snaps up the front. Labeled: Kings Road. Yellow and purple floral print shirt. Labeled: Michael Low, shown with beige and white plaid pants. ($55-60 jacket; $24-28 shirt; $22-24 pants)

Brown and tan "checker board" knit shirt shown with brown pants. ($35-40 shirt; $22-24 pants)

Opposite page:
Green, long-sleeved jacket with shorts. Labeled: Butte Knit. *Modeled by Beth Geronikos.* ($45-50 set)

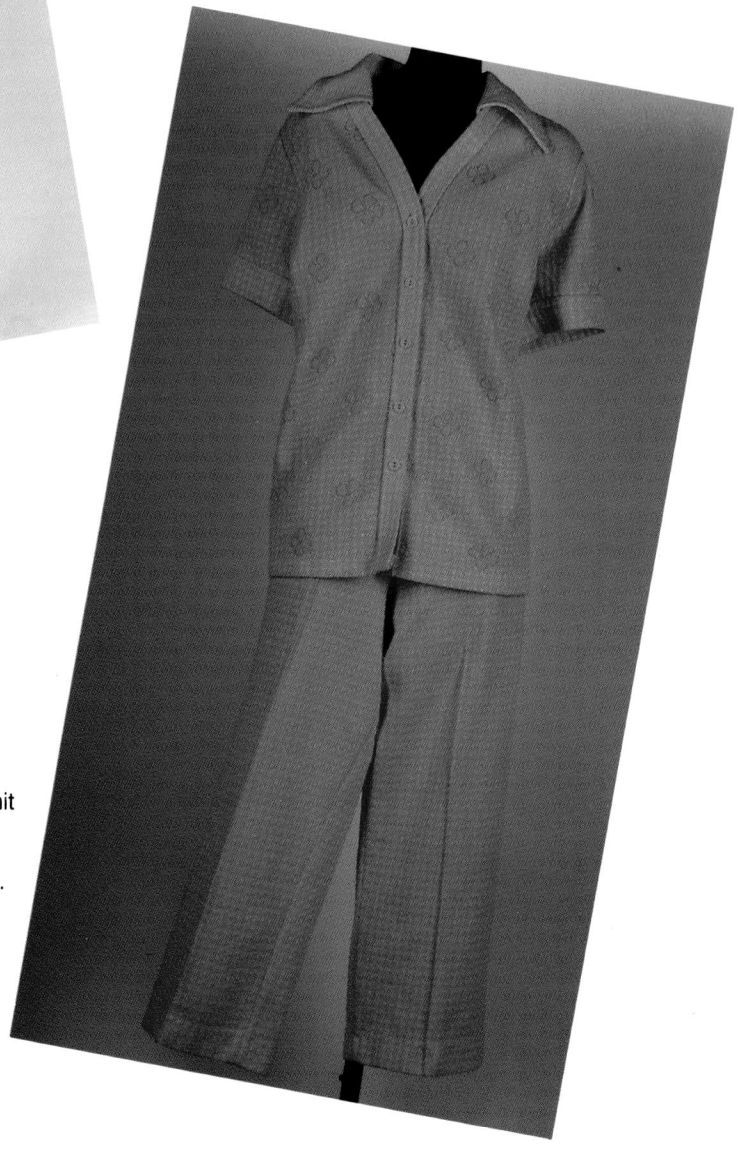

Pants suit with yellow "gingham" knit pants and short-sleeved shirt additionally printed with a yellow floral pattern. Labeled: Gregg's Girl. ($45-50 set)

Opposite page:
Short-sleeved, aqua pants suit with white stitching. *Modeled by Katherine Wellington*. ($55- 65)

Navy blue and red, flared-leg jumpsuit with red zippers at the sleeves and front. Labeled: Charisma. ($85-95)

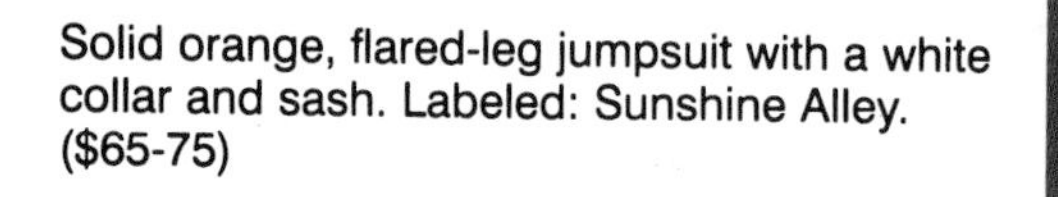

Solid orange, flared-leg jumpsuit with a white collar and sash. Labeled: Sunshine Alley. ($65-75)

Two-piece evening outfit: solid black vest with silver metallic collar and long-sleeves, silver metallic flared-leg pants. ($75-80)

Black, halter top, backless jumpsuit with open sides and palazzo pants. ($85-95)

Coral and peach, short-sleeved, zipper-front, straight-leg jumpsuit. ($65-70)

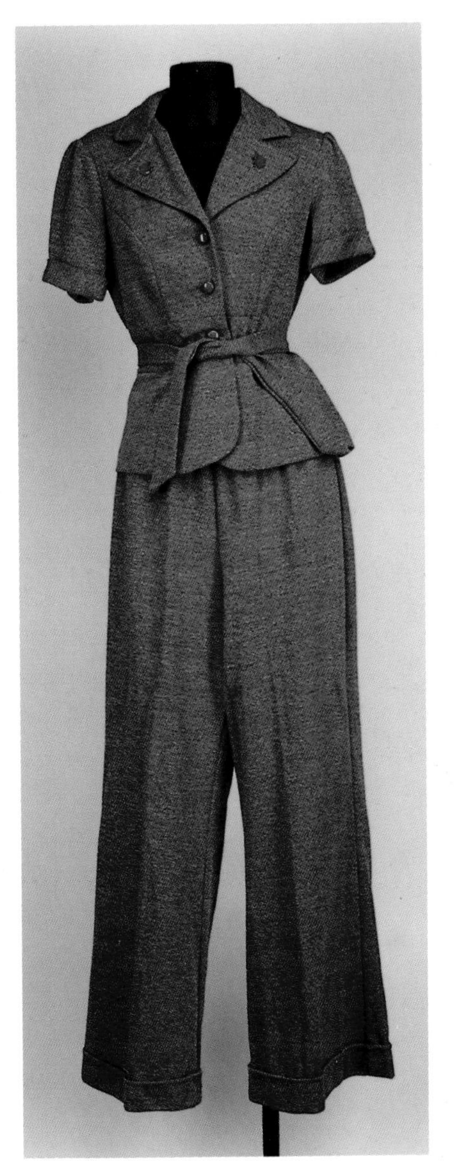

Mauve knit, short-sleeved pants suit with roses appliquéd on the lapels and a tie-belt. ($55-65)

Opposite page:
Mandarin orange, sleeveless jumpsuit with a hood and multi-colored, zig-zag designs across the front. *Modeled by Kayla Agran.* ($95-120)

Teal blue double-knit jacket with three patch pockets, shown with creme and blue print shirt. Jacket labeled: King's Road, Sears. ($45-50 jacket, $22-24 shirt)

Black double-knit suit shown with a white knit shirt with elaborate white stitching. ($75-85 suit; $22-24 shirt)

White double-knit jacket with two patch pockets, shown with a brown and white geometric print shirt. Jacket labeled: King's Road, Sears. Shirt labeled: Mojave of California. ($45-55 jacket; $24-28 shirt)

Black sleeveless, flared-leg jumpsuit with red diagonal zippers. ($90-95)

Light blue jacket with patch-pockets. Labeled: Sears. Navy blue and gray striped shirt. Labeled: Wayne Rogers. Shown with navy blue pants. ($45-50 jacket; $15-18 shirt; $22-24 pants)

Navy blue jacket with patch-pockets. Labeled: Sears. Red, white, and blue Van Knit shirt by Van Heusen, shown with navy blue pants. ($40-45 jacket; $15-18 shirt; $22-24 pants)

Black knit jacket with a "raised diamonds" print, over a red and yellow "zig-zag " print shirt. Labeled: Larry Kane puritan sportswear, shown with black pants. ($50-55 jacket; $22-24 shirt; $22-24 pants)

Gray knit jacket with four patch-pockets, over a gray, red, and black striped shirt, shown with black pants. ($70-75 jacket; $20-24 shirt; $22-24 pants)

Pale yellow jacket with green stitching on the lapels and patch-pockets. Labeled: Don Richards. Maroon, green, and white "fireworks" print shirt. Labeled: Fruit of the Loom, shown with black pants. ($65-70 jacket; $20-24 shirt; $22-24 pants)

Mandarin orange pants suit with white stitching and flared-leg pants. Labeled: White Stag. ($45-50)

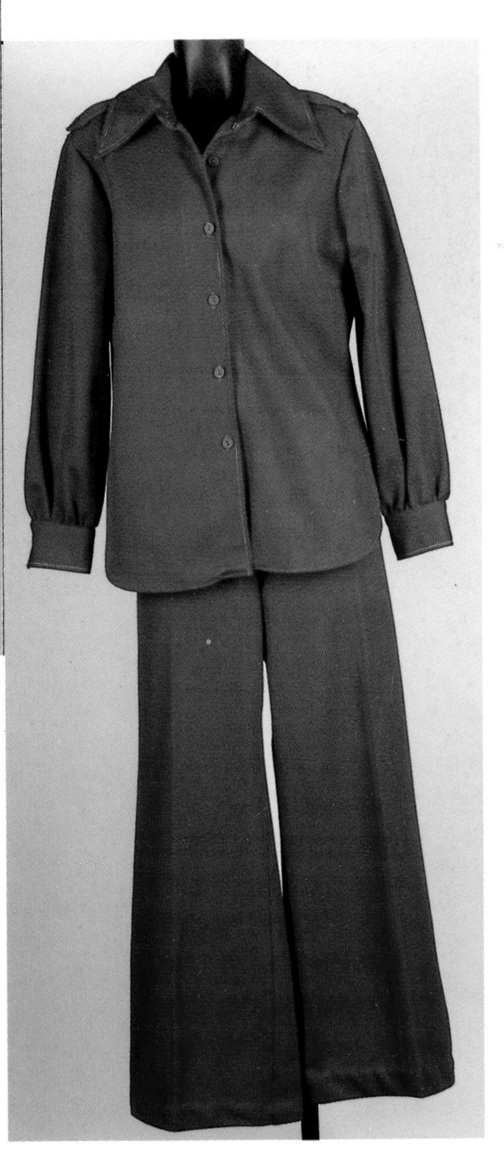

Index of Labeled Clothing

INJUSTICE

GODS AMONG US: YEAR TWO

VOLUME 2

INJU

GODS AMON

Tom Taylor Marguerite Bennett
Writers

Bruno Redondo Thomas Derenick Mike S. Miller
Julien Hugonnard-Bert Vicente Cifuentes Daniel HDR
David Yardin Jheremy Raapack Xermanico Juan Albarran
Artists

Rex Lokus J. Nanjan (NS Studios)
Colorists

Wes Abbott
Letterer

Jheremy Raapack & David Lopez and Santi Casas of Ikari Studio
Cover Artists

STICE

US: YEAR TWO

VOLUME 2

Aniz Ansari Assistant Editor – Original Series
Jim Chadwick Editor – Original Series
Jeb Woodard Group Editor – Collected Editions
Rachel Pinnelas Editor – Collected Edition
Louis Prandi Publication Design

Bob Harras Senior VP – Editor-in-Chief, DC Comics

Diane Nelson President
Dan DiDio and Jim Lee Co-Publishers
Geoff Johns Chief Creative Officer
Amit Desai Senior VP – Marketing & Global Franchise Management
Nairi Gardiner Senior VP – Finance
Sam Ades VP – Digital Marketing

INJUSTICE: GODS AMONG US: YEAR TWO VOLUME 2

"The Quiver" Bruno Redondo Julien Hugonnard-Bert Artists
"Resistance" Thomas Derenick Mike S. Miller Artists Rex Lokus Colorist
Cover Art by Stephane Roux

I NEED YOU HERE, DINAH. I NEED YOUR LEADERSHIP ON THE GROUND.
I'LL ONLY BE GONE A DAY. COMMISSIONER GORDON CAN COORDINATE THINGS ON THE GROUND IN GOTHAM UNTIL I GET BACK. HE'S FAR MORE QUALIFIED TO LEAD THE RESISTANCE THAN I AM ANYWAY.
I DON'T THINK THAT'S TRUE.
REALLY?
REALLY.
SO IT'S NOT JUST THAT YOU'RE HAVING A HARD TIME TELLING YOUR DAD WHAT TO DO, AND YOU'RE FAR HAPPIER ORDERING ME AROUND?
THAT'S... ANNOYINGLY INSIGHTFUL.
DON'T WORRY. I'LL BE BACK TOMORROW AND READY TO BLINDLY DO YOUR BIDDING AGAIN.
I APPRECIATE THAT.
I JUST HAVE A FEW OF OLLIE'S AFFAIRS TO SEE TO. AND I'VE BEEN PUTTING OFF COMING BACK TO THE ARROW CAVE.
'ARROW CAVE' IS A TERRIBLE NAME.
HARLEY QUINN SAID HE SHOULD CALL IT 'THE QUIVER.'
THAT IS ACTUALLY BETTER.
I KNOW.
ORACLE--

THE QUIVER

--I'M GOING TO HAVE TO CALL YOU BACK.

WHAT ARE YOU DOING HERE?
UM... PRACTICING ARCHERY IN COMFORTABLE SLIPPERS?
WHAT ARE YOU DOING *HERE?*
I...I DIDN'T KNOW WHERE ELSE TO GO. GREEN ARROW SAID--

I WANT YOU TO LEAVE.
YOU CAN'T TEL ME WHAT TO DO. TH ISN'T YOU CAVE!

YOU *REALLY* WANT TO DO THIS?

FLUFFY SLIPPER ATTACK!
THP
THP
WOW, BLACK CANARY. YOU LOOK PRETTY CALM FOR SOMEONE WHO'S JUST CAUGHT EXPLODING SLIPPERS.
THEY'RE NOT REALLY EXPLODING SLIPPERS, ARE THEY?
OF COURSE NOT. THAT WOULD BE CRAZY.

SCREEEEEEEEEE
SCREEEE
OW!
OH, YOU'RE GETTING SUCH A GIANT-MALLET-SQUISHING FOR THAT.
HAAAAAARGH!
WAIT!

I THOUGHT WE WERE FIGHTING.
WE ARE. JUST...GIVE ME A SECOND.
IS THAT ALLOWED?
HEY, ARE YOU OKAY?
HAND ME THAT BUCKET.
WHAT?
NOW!
WHY DO YOU NEED A--
HURGHH!

EWW.
WHAT'S WRONG WITH YOU?
IT'S... I'M FINE!
WAIT.
WAIT!
IS IT MORNING SICKNESS?
I AM NOT DISCUSSING THIS WITH YOU. LET'S JUST GET BACK TO ME THROWING YOUR ASS OUT OF HERE.
IT IS! YOU'RE PREGGERS!
I SAID I'M NOT--
HURGHHH!
HA!

YOU THINK THIS IS FUNNY?!
I THINK IT'S HILARIOUS!
OH, COME ON. HOW IS IT NOT FUNNY? YOU'RE A SUPERHERO WITH MORNING SICKNESS.
YOU SAVE THE WORLD. AND THEN THROW UP.
AND... OH, MY GOD.
WHAT?
I'M PICTURING YOU IN THE THIRD TRIMESTER. ALL FAT BUT STILL WEARING THE FISHNETS.
STOP IT.
I CAN'T. THAT IMAGE IS SEARED FOREVER INTO MY BRAIN.
CAN WE PLEASE GET BACK TO BEATING EACH OTHER UP? IT'S FAR LESS PAINFUL.
OH, DON'T BE STUPID. I'M NOT GOING TO FIGHT YOU NOW THAT I KNOW YOU'RE PREGNANT.
I REMEMBER HOW HARD IT IS.

YOU REMEMBER...?
HUH?
YOU REMEMBER WHAT?
HARLEY? ARE YOU...? DID YOU HAVE A CHILD?
LET'S GO BACK TO HITTING EACH OTHER. THAT WAS FUN.
HARLEEN?

I...I HAVE A DAUGHTER. HER NAME IS LUCY.
SHE LIVES WITH MY SISTER.
SHE'S FOUR AND SHE LIKES TUTUS AND TOY CARS.
SHE ESPECIALLY LIKES PUTTING TUTUS ON TOY CARS. SHE LIKES THE IDEA OF A BALLERINA CAR.
I THINK THAT MAKES PERFECT SENSE.
MISTAH J DIDN'T KNOW ABOUT HER. I KNEW THAT WOULDN'T...THAT HE WOULDN'T...
HIS WORK WAS TOO IMPORTANT. HE COULDN'T HAVE A DAUGHTER.
I WENT AWAY FOR ALMOST A YEAR.
WHEN I CAME BACK TO MY PUDDIN', HE WAS SO FUNNY.
HE ACTED LIKE HE DIDN'T EVEN NOTICE I WAS GONE.
GOOD JOKE, HUH? GONE A WHOLE EAR AND HE PRETENDS LIKE HE DIDN'T EVEN CARE?
IT WOULDN'T HAVE WORKED. IT'S BETTER THIS WAY. LUCY'S BETTER OFF.
SHE'LL HAVE A NORMAL LIFE FILLED WITH NORMAL DAYS AND DANCING CARS.

IS THE BABY OLIVER'S?
YOU'RE NOT SUPPOSED TO KNOW HIS SECRET IDENTITY... AND OF COURSE IT'S OLLIE'S!
I SAW THE NEWS. OLIVER DIED.
HE WAS... HE WAS ONE OF THE GOOD ONES. WASN'T HE?
OLLIE BELIEVED IN YOU, YOU KNOW? I HAVE NO IDEA WHY. THAT MAN COULD HOLD A GRUDGE LIKE NO ONE ELSE BUT HE SAW SOMETHING IN YOU.
SOMETHING NOT BEYOND REDEMPTION.
DID SUPERMAN KILL HIM?
YES.
WHAT ARE YOU GOING TO DO?
I'M GOING TO MAKE SURE SUPERMAN CAN'T HURT ANYONE ELSE EVER AGAIN.
I'M GOING TO HELP YOU.
I DON'T NEED *YOUR* HELP.
SUPERMAN KILLED MY MAN, TOO, YOU KNOW? I HAVE AS MUCH REASON TO HATE HIM AS ANYONE.
AND OF COURSE YOU NEED MY HELP. I CAN DO THINGS YOU CAN'T.
LIKE WHAT?
LIKE FIGHT WITHOUT NEEDING A BUCKET NEARBY.
PLEASE.
LET ME HELP.

RESISTANCE
HOW IS THIS POSSIBLE?
I DO NOT KNOW. IT'S AS IF EARTH HAS SIMPLY DISAPPEARED.
AND THE LANTERNS WE SENT TO APPREHEND SUPERMAN?
STILL NO WORD.
HOW CAN AN ENTIRE WORLD BE REMOVED FROM OUR SIGHT?
I THINK I MIGHT BE ABLE TO FILL YOU IN.

GUY GARDNER? WHAT HAPPENED?
I WAS ON MY WAY TO SEE YOU WHEN I GOT JUMPED BY TWO OF THE SINESTRO CORPS. I WAS ABLE TO OVERCOME ONE--SEVERELY PISS OFF THE OTHER--AND ESCAPE.
THEY WERE SCOUTS. SENT WATCH THE W BETWEEN OA EARTH.
WHY WOULD THEY BE WATCHING THAT ROUTE?
THAT'S WHY I WAS ON MY WAY HERE. IT'S BECAUSE SINESTRO IS ON EARTH.
HE'S JOINED FORCES WITH SUPERMAN.
THIS IS AN UNFORESEEN AND TROUBLING DEVELOPMENT.
HOWEVER, IT EXPLAINS MUCH. SINESTRO HAS ALWAYS HAD WAYS OF MASKING HIS PRESENCE FROM OUR SIGHT.
WHAT OF THE LANTERNS WE SENT?
I DON'T KNOW. MOST OF THEM WERE ALIVE WHEN I LEFT BUT I GOT OUT OF THERE PRETTY QUICKLY.
BUT... I GUESS WE WORKED OUT WHAT HAPPENED TO KYLE.

AND HAL JORDAN AND JOHN STEWART?
I DON'T KNOW.
I CAN'T SEE EITHER OF THEM GOING ALONG WITH SINESTRO. BUT...I'VE BEEN WRONG ABOUT A LOT OF FRIENDS LATELY.
DESPITE THEIR INSUBORDINATION, I AM WILLING TO TRUST THEM BOTH FOR NOW. THEY COULD STILL PROVE USEFUL WHEN THE TIME COMES.
WHAT DO YOU MEAN *'WHEN THE TIME COMES?'*
THE TIME HAS TO COME NOW, GANTHET. WE CAN'T LEAVE SINESTRO--
THERE IS NOTHING WE CAN DO FOR NOW. TO TAKE ON SUPERMAN'S FORCES AND THE SINESTRO CORPS TOGETHER WOULD TAKE A HUGE FORCE. WE CAN'T RECALL ALL OF THE LANTERNS AT THIS TIME. THERE ARE OTHER WARS AND OTHER PROBLEMS ELSEWHERE THAT CAN'T BE IGNORED.
WHAT ARE YOU SAYING?
EVEN IF WE HAD THE ATTACK FORCE, TO ATTACK EARTH COULD RESULT IN THE DESTRUCTION OF THE VERY PLANET WE'D BE FIGHTING TO PROTECT.
SO, WHAT THEN? THE PLAN OF THE ALL-WISE GUARDIANS IS TO LEAVE EVIL-DOUCHE SUPERMAN ALONE TO PLOT WITH HIS NEW SUPER-BEST-FRIEND SINESTRO?
NO. BUT TO ACT RASHLY, TO ACT HASTILY, WOULD BE TO PLAY DIRECTLY INTO SINESTRO'S HANDS. WE WILL FIND A WAY TO BRING SUPERMAN--AND SINESTRO--TO JUSTICE. BUT THE EARTHLINGS ARE ON THEIR OWN FOR NOW.
NO--

"--THEY'RE NOT COMPLETELY ALONE."
GOTHAM.
HOW MANY, ORACLE?
THERE SHOULD ONLY BE TWO.
AND THEY SHOULD BE ISOLATED, BLACK CANARY, WELL AWAY FROM THE REST OF THE FORCE. THEY'RE CLEARLY GROWING OVERCONFIDENT WITH THEIR POWER.
WELL, THAT WORKS FOR US.
YOU EVER GOING TO TELL US HOW THIS MYSTERIOUS 'ORACLE' KNOWS THE PATROL DETAILS?
NOPE.
AGHHHH!
THAT'S YOUR CUE, DETECTIVES.
BULLOCK. MONTOYA. REMEMBER, SHOW NO SIGNS OF ENHANCED ABILITIES UNTIL WE'VE TAKEN OUT THEIR COMMUNICATIONS. THEY CAN'T BE ALLOWED TO REPORT OUR CAPABILITIES BACK TO SUPERMAN.

WHAT PART OF 'CURFEW' DO YOU PEOPLE NOT UNDERSTAND? YOU THINK WE WANT TO MAKE AN EXAMPLE OUT OF YOU IN FRONT OF YOUR FRIENDS? YOU THINK WE WANT TO HURT YOU?
CRCK
PLEASE... PLEASE...
NAAARGH!
BANG
LET THEM GO.
HAVEN'T YOU HEARD? BULLETS DON'T HURT--
BANG
BULLETS MAY NOT HURT YOU, BUT SHOOTING YOU IN THE FACE STILL FEELS PRETTY DAMN GOOD.
THD

YOU? WHAT ARE YOU DOING STANDING? DIDN'T I HIT YOU THROUGH A WINDOW?
THAT WAS YOU?
HARVEY. NOT YET.
THIS GUY MADE ME TAKE OFF MY HAT IN PUBLIC.
YOU'RE A COP, RIGHT? WHAT IS THIS?
THE GOTHAM CITY POLICE DEPARTMENT LOOKING TO GET INTO SOME TROUBLE?
WELL?
HANG ON. I'LL ANSWER THAT QUESTION AS SOON AS MY FRIEND LANDS.

HUH?
CHZZZT
CHZZZT
THEIR COMMUNICATIONS ARE OUT.
THE WAREHOUSE DIRECTLY IN FRONT OF YOU IS ABANDONED. TAKE THEM OFF THE STREET.
WITH PLEASURE.
TO ANSWER YOUR QUESTION. YES, WE'RE LOOKING TO GET INTO SOME TROUBLE.

TOOOOM
CRNCH
WELL, THAT WAS SATISFYING.
HOW THE HELL...?
WE CAN STILL TAKE YOU.
MAYBE. MAYBE YOU COULD TAKE US TWO-ON-TWO IN A FAIR FIGHT--
--BUT WE'RE WAY PAST A FAIR FIGHT.

CSHH
CSHH
CSHH
HOW ABOUT YOU? LOOKING TO GET INTO SOME TROUBLE?
RUNNING AWAY? REALLY?
WORST.
SUPERSOLDIER.
EVER.
CRNCH
"COMING FAST. WEST WALL."

CRCK
HNNF!
YOU HAVE THE RIGHT TO REMAIN SILENT. ANYTHING YOU SAY--
NO.
COMMISH?
PICK HIM UP. AND BRING THE OTHER ONE.
THEY'RE AN INVADING ARMY, NOT CRIMINALS.
WE'RE GOING TO BE INTERROGATING THEM FOR AS MUCH INFORMATION AS WE CAN GET, AND HOLDING THEM WITHOUT DUE PROCESS.
WE DON'T READ THEM THEIR RIGHTS. THEY DON'T HAVE ANY.

Marguerite Bennett & Tom Taylor Writers

"Closing Time" **Vicente Cifuentes Daniel HDR Mike S. Miller David Yardin** Artists **Rex Lokus** Colorist

"The Ur-Forge - An Untold Injustice Tale" **Jheremy Raapack** Artist **David Lopez & Santi Casas of Ikari Studio** Colorists

Cover Art by **Jheremy Raapack & Mark Roberts**

RAAPACK 2014

HE HEADQUARTERS
F THE RESISTANCE,
OTHAM CITY.

RARE RESPITE FROM EVENTS
THE WORLD OUTSIDE.

THE NEXT ONE, THE NEXT ONE!

OH, JEEZ, THIS IS MY MOM AND DAD--SHE JUST MADE DETECTIVE--LOOK AT THE *HAIR*--

THE DAY I GOT MY LICENSE--MY DAD WOULD WAIT UP FOR ME EVERY NIGHT, AND I MEAN *EVERY NIGHT*--
THE NEXT ONE!

I...
SORRY, I... THERE'S SOME WORK I HAVE TO GET DONE.

CLOSING TIME
LATER
SOMEONE OUT AFTER CURFEW--?
DAD? DAD?!
BARBARA, WHAT IN THE NAME OF--
NO-- GOD, NO--

YOUR MOTHER...
...VANISHED FIFTEEN YEARS AGO.

THE FOLLOWING NIGHT.
DAD, DON'T-- DON'T USE THOSE DAMN THINGS--
I DON'T CARE WHAT ENHANCEMENTS THEY HAVE, THEY'RE SPEEDING UP YOUR CANCER--
BARBARA... WHEN YOUR MOTHER DISAPPEARED...
SHE'S DEAD, DAD.
THEY NEVER FOUND A BODY.
WE'VE LOST SO MUCH. I CAN'T LOSE YOU TOO. NOT NOW, NOT OVER THIS.
SOMETIMES... THE THINGS WE LOVE COME BACK TO US.

WHEN WE'RE ON PATROL, I'M ORACLE, DAD.
THIS ISN'T MY FIRST RODEO.
OR MY... FIRST TURN AT CRIMINAL SABOTAGE, COME TO THINK OF IT.
WE'VE GOT ANOTHER MARCH COMING UP ON HAYWARD STREET--DODGE INTO THE ALLEY NEAR MIDLAND--
NO, HEAD BACK. YOU'RE A STREET TOO FAR!
HERE WE GO, SON--
IT'S ALL RIGHT, YOU'LL WAKE UP WITH NOTHING WORSE THAN A CASE OF THE BROWN BOTTLE FLU.
THERE'S A GOOD MAN.

DO YOU HAVE VISUAL OF HER?
I'M INTO THE SECURITY CAMERAS, REDIRECTING THEIR FEEDS HERE INSTEAD OF TO THE MILITARY BASE...
BANK LEFT. I'VE GOT HER REAPPEARING BY THE...
BY THE *WHERE*, BARBARA?
BY THE... BOTANICAL GARDENS.
GOTHAM CITY BOTANICAL GARD
BARBARA...

WAIT...
PLEASE, I--
...CAN'T BE...

BARBARA...
HOW IS THIS...IS THIS SOME TRICK...?
I'M SO SORRY. I'M SO SORRY, I HAD TO HIDE...
YOU NEVER CAME HOME...YOU WERE ON PATROL, AND YOU NEVER...WE HUNTED FOR MONTHS, WE WAITED YEARS, AND YOU--
IT DOESN'T MATTER, IT-- OH!

NO...!
NOT BARBARA...
BASIL KARLO.

CLAYFACE.
GORDON-- DETECTIVE GORDON, STOP, PLEASE--
PLEASE! THEY'RE GOING TO KILL ME!
I ESCAPED THE PURGE IN ARKHAM-- PLEASE, I BEEN LOOKING AND LOOKING FOR YOU. I'M BEGGING YOU, TAKE ME IN!
WHATEVER YOU WANT, I'LL DO, I'LL BE--
THE OLD GCPD, THEY'D THROW ME ONTO A CHAIR, INTO A CELL, BUT THESE SOLDIERS WILL KILL ME, GORDON, PLEASE!
GIVE ME ONE DAMN REASON IN THIS WHOLE WRETCHED WORLD I SHOULD HELP A PIECE OF POND SCUM LIKE YOU.
BECAUSE...
I CAN GIVE YOU THE ONE FIXED, PERMANENT THING IN THIS MESSY, MUTABLE WORLD...
I CAN GIVE YOU CLOSURE.
I CAN TELL YOU... WHERE THEY BURIED YOUR WIFE.

"MY WIFE IS...
"...DEAD."
"I DIDN'T MEAN FOR THIS TO GO SO FAR, ME LOOKING LIKE YOUR WOMAN, GORDON... I WAS JUST TRYIN' TO GET YOUR ATTENTION...
"SAW THE PICTURE OF YOU AND HER ON YOUR DESK YEARS AGO, FIRST TIME THE GCPD BROUGHT ME IN.
"I WAS A DUMB KID... RUNNING WITH A BUNCH OF JACKALS WHO THOUGHT MY PARTY TRICK WAS *NEAT.*"
FREEZE! HANDS IN THE AIR!
"YOUR WOMAN, SHE WAS HUNTING THE RED HOOD GANG, WASN'T SHE?
"WRONG PLACE. WRONG TIME. WRONG GANG.
"BLOOD ON THE AIR... THEY WAS NO BETTER THAN JACKALS, AFTER ALL.
"SHE DIED IN THE LINE OF DUTY, GORDON.
"I'LL... I'LL TAKE YOU, TO PAY YOUR RESPECTS."

THE PIER. NOW.
DOWN HERE.
B--ORACLE, YOU LISTENING? WATCHING?
I-I'M HERE.
YOU SEE ANY SIGN OF A TRAP?
NO, DAD.
I'M SORRY, GORDON.
IT'LL BE ALL RIGHT, BARBARA...
WHEN THE TIME COMES... YOUR MOTHER WILL... YOUR MOTHER WILL BE WAITING FOR ME.
I'LL BE SEEING HER AGAIN.

COMMISSIONER!
WHAT THE HELL ARE YOU THINKING, GORDON?
DAD.
I'VE GOT BULLOCK...I'VE GOT A DETACHMENT DOWN AT THE PIER, BARBARA. THEY'LL...BRING YOUR MOTHER'S BODY BACK.
WE'RE BRINGING HER HOME.
GORDON. WE WANT ANSWERS.
TO WHICH YOU'RE ENTITLED.
MR. BASIL KARLO HERE WILL BE EXCHANGING HIS SERVICES FOR PROTECTION FROM THE MILITARY POLICE. AFTER THE LOSS OF THE MARTIAN MANHUNTER, I BELIEVE HE'LL BE OF GREAT ASSISTANCE TO YOU.
HARLEY. WITH ME.
OH, GOODIE, SECRET CONSPIRACIES--

YOU'RE GOING TO BEGIN TRAINING WITH THE BIRDS OF PREY. ONE LUMP OUT OF THE LINE AND BACK IN THAT GARDEN AS *FERTILIZER*, KARLO.
YOU DON'T REALLY MEAN THAT, GORDON, DO YOU--?
OUCH!
NOT SURE WHAT GOOD I'M SUPPOSED TO BE-- I NEVER MET SUPERMAN OR ANY OF HIS CRONIES. SHIFTING DOESN'T MEAN I TAKE ON ANY POWERS, YOU KNOW--
YOU'LL EARN YOUR KEEP, KARLO.
COME ON, KARLO. WE'RE KEEPING YOU ON ICE UNTIL WE SEE THE RESULTS.
MORE THAN ONE WAY TO GET CLOSURE, MISTAH G.

WHO...?
DICK...?

STOP!
I TOLD YOU TO STOP.
I DID-- OH...
HOW... DARE YOU.
I-I'M REAL SORRY--DIDN'T MEAN ANY HARM-- I READ LOCAL OBITS, I PICK FACES NO ONE'S USING--
WE DON'T EVEN NEED TO GO INTO WHY YOU'RE OUT OF YOUR CELL. YOUR STORY HAS BEEN WRONG FROM THE BEGINNING.
WANT TO KNOW WHAT I FOUND IN YOUR DNA? KRYPTONIAN BLOOD. YOU'RE A LIAR.
YOU'VE MET SUPERMAN.

SUPERMAN SENT YOU...TOLD YOU EXACTLY WHO AND WHERE TO BE TO DRAW US OUT.
HE WANTED US TO TAKE YOU IN...WANTED YOU TO JOIN US, TORMENT US, TURN US AGAINST EACH OTHER--

HE WAS PRETTY SPECIFIC ABOUT THAT, YEAH.

"LET THEM SEE THE FACES OF THEIR DEAD," HE SAYS TO ME. "LET THEM SEE AND NEVER FORGET...
"...AS I **MUST** ALWAYS SEE AND **CAN** NEVER FORGET."

RAAAAAA!

ONE WAY OR ANOTHER--
WHMP

IT'S ALWAYS BACK INTO THE DIRT AND FILTH WITH YOU.

WAS THAT EVEN MY MOTHER'S BODY, DOWN UNDER THE PIER?!
OH, THAT'S YOUR MOMMY.
ONLY I WASN'T JUST STANDING PRETTY WHEN IT HAPPENED.
RAAARG
BARBARA!
RRRRRG!
UNNH!
THWCH

WHAM
SCREEEE
WHUMP

LATER...
HARLEY, I'M NOT HURT THERE--
SHH.
THEY'VE GOT *HELLO KITTY* ON THEM. THEY'RE FASHIONABLE.
GOOD AS NEW.
OH, DAD...
AND WHAT ARE WE DOING WITH THIS GUTLESS PIECE OF--
I'VE GOT AN IDEA!

SOON...
IT WON'T... KILL HIM--*OOF*, THIS IS A TIGHT FIT--BUT HE'LL--*OOF*--STAY OUTTA TROUBLE.

SEE?

CLOSURE.

BARBARA?
HERE, DAD.
YOU CAN TURN IT BACK ON, IF YOU WANT. I DIDN'T MEAN TO INTERRUPT.
I KNOW.
IT'S JUST SO HARD, DAD. KNOWING THAT IN A MONTH, A WEEK, A DAY... THIS IS ALL I'M GOING TO HAVE LEFT OF YOU.
PICTURES. COLD CASES. GREATEST HITS.
KNOWING THIS MOMENT IS ONE OF THE LAST THAT WE'LL EVER BE TOGETHER.
I WISH...
I WISH I COULD STAY.

I NEVER MEANT TO LEAVE YOU ALONE, BARBARA...
I'LL FIND A WAY TO MAKE MY PEACE...
I'LL FIND A WAY TO GO ON.
I'LL NEVER BE ALONE.
I'LL KNOW YOU'LL BE WAITING UP FOR ME WHEN I GET IN...
NO MATTER HOW LATE I LEAVE...
...NO MATTER HOW LATE I COME BACK HOME.
I KNOW, DAD...
...YOU'LL ALWAYS BE WAITING THERE FOR ME.
END

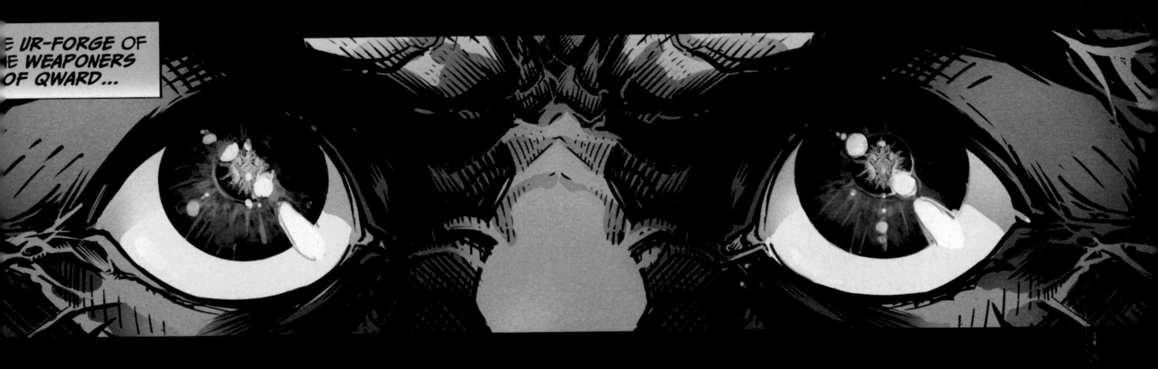

RIMAL AND ANCIENT...
OOT OF VIOLENCE,
WOMB OF DEATH...

EARTH
THE UR-FORG
An Untold Injustice Tal
THOSE BLIND, MANIPULATIVE FRAUDS.
YOU ARE CERTAIN THE GUARDIANS HAVE LOCATED THE UR-FORGE OF THE WEAPONERS? THE FORGE IS THOUGHT TO BE BUT A MYTH BY MOST LANTERNS...
I'VE SEEN IT. I HAVE THE COORDINATES FOR YOU.
PLEASE. I AM ASKING YOU TO DO THIS FOR THE PROTECTION OF EARTH AND THE REST OF THIS ENTIRE GALAXY.
THE GUARDIANS ARE HUNTING THE FORGE, IF THEY HAVE NOT FOUND IT ALREADY. WHAT IT COULD UNLEASH ON ALL OF US IS...
UNIMAGINABLE?
HAL, FOR YOUR KNOWLEDGE OF THE GUARDIANS, AND SINESTRO, FOR YOUR KNOWLEDGE OF THE WEAPONERS, I MUST ASK YOU TO DO THIS.
FOR THE SAKE OF OUR PEOPLE--OUR PLANET--
BRING ME THE UR-FORGE.

SUPERMAN'S COORDINATES DID NOT LEAD TO QWARD ITSELF--
--BUT TO A DISTANT MOON, THE CRADLE OF LIFE, PERHAPS, FOR THE WEAPONERS--
--THE BIRTHPLACE OF THEIR FIRST CREATIONS.

THIS PLACE...
THE EVOLUTIONARY CESSPOOL FROM WHICH THE WEAPONERS CREPT.
THEIR DESIRE FOR CRAFT, FOR STRENGTH, FOR SUPERIORITY, LED THEM TO STRIP THEIR WORLD, REPLACING EACH ASPECT.
BUT THIS WAS MILLENNIA BEFORE THEY SLIPPED THROUGH THE TEAR IN THE UNIVERSE.
THE COURSE THEY WOULD HAVE TAKEN, HAD THEY REMAINED ON THIS LITTLE ROCK... ONE CANNOT EVEN IMAGINE. THEY MAY HAVE EVEN BECOME PEACEFUL, IN TIME
"AT SOME CROSSROADS, ONE MAY LOOK BACK, TRYING TO GATHER WHICH PATH IT WAS THAT LED ONE TO COME TO BE WHERE ONE IS, BUT OTHER ROADS... THERE HAVE BEEN TOO MANY DIVERGENCES.
"IT IS OFTEN IMPOSSIBLE TO SAY WITH CERTAINTY ON WHAT CHOICES OR CHANCES A LIFE HANGS."
SKREEEEEE!!
JORDAN, LET ME--
RAT-TAT-TAT
BACK OFF.
AFTER ALL WE HAVE BEEN THROUGH, STILL...?
AFTER ALL WE'VE BEEN THROUGH, YES.

LATER.
WE'LL FOLLOW THE SIGNAL IN THE MORNING. THERE'S JUST TOO MUCH INTERFERENCE FROM THESE MECHS RIGHT NOW, BUT NIGHT SHOULDN'T LAST LONG ON A MOON THIS SMALL.
SHALL I TAKE WATCH--?
NO.
YOU DON'T TRUST ME. YOU DON'T NEED TO.
YOU TRUST *HIM*.
SUPERMAN. HE CHOSE YOU FOR THIS MISSION, AND HIS JUDGMENT...
KAL-EL IS MY FRIEND.
I COULD ASK YOU WHAT HAPPENED BETWEEN US, HAL. HOW WE FELL SO FAR. YET I THINK ALL IT WOULD DO IS HIGHLIGHT THAT WE HAVE FALLEN SO MUCH FARTHER STILL.
WOULD YOU EVER HAVE GUESSED YOU WOULD LONG FOR THE DAY WHEN WE HACKED AND SLASHED AT ONE ANOTHER LIKE CIVILIZED KILLERS?
LANTERNS AREN'T KILLERS.
SO IT WAS SAID OF SUPERMAN, TOO.
I HAVE SEEN THIS ALL BEFORE, HAL... YOU DO NOT WANT TO PUT A FINGER ON THE MATTER, DON'T WANT TO SAY THE WORD ALOUD, BUT THAT IS WHAT IS TROUBLING YOU, BENEATH THE PRACTICALITY OF THIS MISSION.
I HAVE SEEN WORLDS DIE BEFORE, BECAUSE ONE MAN THOUGHT HIMSELF A SAVIOR.
I WAS THAT ONE MAN. AND I BROUGHT DEVASTATION TO MY WORLD.
I WAS YOUR FRIEND TOO, ONCE.

AND I THINK, WHEN YOU SEE KAL... IT IS ME YOU TRULY SEE.
ME, WHEN YOU FAILED TO STOP ME. ME, GIVEN A SECOND CHANCE.
AND THIS TIME, YOU MIGHT SAVE SO MANY LIVES.
ARE YOU GOING TO MAKE THE BREAKFAST IN THE MORNING OR SHALL I?
SENSE OF HUMOR COME IN THE GENOCIDAL STARTER KIT?
I REMIND YOU, HAL, I HAVE SEEN MANY WORLDS DIE. GALLOWS HUMOR COMES TO US ALL, IN THE END.
UNLESS...
UNLESS WHAT?
YOU KNOW UNLESS WHAT.
UNLESS SOMEONE DOES SOMETHING. UNLESS THE WORLD DOES NOT DIE.
THE WEAPONERS TORE THEIR WORLD APART, REPLACED IT WITH SOMETHING UNRECOGNIZABLE, MECHANICAL, AND OBSCENE, RATHER THAN SEE IT DIE...

...AND THEY DID NOT STOP WHEN THEY HAD PICKED THEIR WORLD CLEAN.
SKREEEEE!!
IT MUST'VE SENSED THE CAMPFIRE--
MOVE!
IS IT A MACHINE, HAL, OR IS IT AN ANIMAL? IF IT HAS THE ILLUSION OF CHOICE, IS IT TRULY FREE?
SINESTRO, I'M GONNA NEED LESS SOPHOMORE YEAR PHILOSOPHY AND A LOT MORE "ADVANCED SEMINAR IN RUNNING."
THE UR-FORGE--IT'S NEAR! IT'S RETURNED TO THE RADAR--
WE'RE NOT ANYWHERE NEAR THE COORDINATES SUPERMAN GAVE US--
THE UR-FORGE...HAS MOVED.
WHAT ARE YOU THINKING?

NOT SO FAST, HAL!
WE CAN'T LET IT LOSE US!
SKREEE!!
"IT'S TRYING TO CONVERT US TO MACHINES
"AND IT CAN REPAIR ITSELF INSTANTLY."
JUST-- A MOMENT-- MORE--
SKRUNCH

SKREEEEEEEEEEEEE!!
I BELIEVE WHAT WE SEEK IS STRAIGHT AHEAD.
KRUNNCH
KRACKLE

HI, GUYS.
SKREEEEEE!!

"SUPERMAN SAYS 'HELLO'."

WE GOT IT--WE ACTUALLY GOT IT--
GOD FORGIVE US.
YES. WE DID IT.
GOD DAMN US.
IT WILL TAKE BOTH OF US TO TRANSPORT IT OFF WORLD... YOU'LL NEVER MOVE IT ON YOUR OWN.
IT WILL PASS INTO SUPERMAN'S HANDS, OR SINESTRO'S, OR THE GUARDIANS'.
I DON'T KNOW WHICH IS WORSE.
I JUST DON'T--

YOU... YOU, OF ALL PEOPLE...
NO ONE CAN USE IT NOW.
YOU WERE WORRIED, WEREN'T YOU? THAT LITTLE TANG OF NOT-QUITE-FEAR...
WERE YOU MORE AFRAID OF WHAT THE UR-FORGE WOULD DO IN THE GUARDIANS' HANDS, SUPERMAN'S... OR MINE?
HOW COULD I BE AFRAID...
...WHEN YOU WERE THE ONE AT MY BACK?

ARTH.
HAL? EMPTY-HANDED? THE GUARDIANS HAVEN'T--?!
I'M SORRY, KAL-EL.
THE UR-FORGE WAS DESTROYED. WE COULDN'T SALVAGE IT.
DESTROYED...
THAT... IS FOR THE BEST. OUR PEOPLE WILL BE SAFE AFTER ALL.
THANK YOU, HAL. THANK YOU, SINESTRO.
AT LEAST IT DIDN'T FALL INTO THE WRONG HANDS.
NO...
NO, IT DID NOT.
END

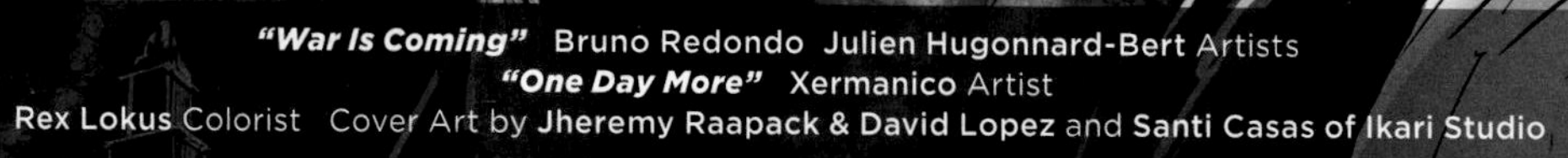
"War Is Coming" **Bruno Redondo** **Julien Hugonnard-Bert** Artists
"One Day More" **Xermanico** Artist
Rex Lokus Colorist Cover Art by **Jheremy Raapack & David Lopez** and **Santi Casas of Ikari Studio**

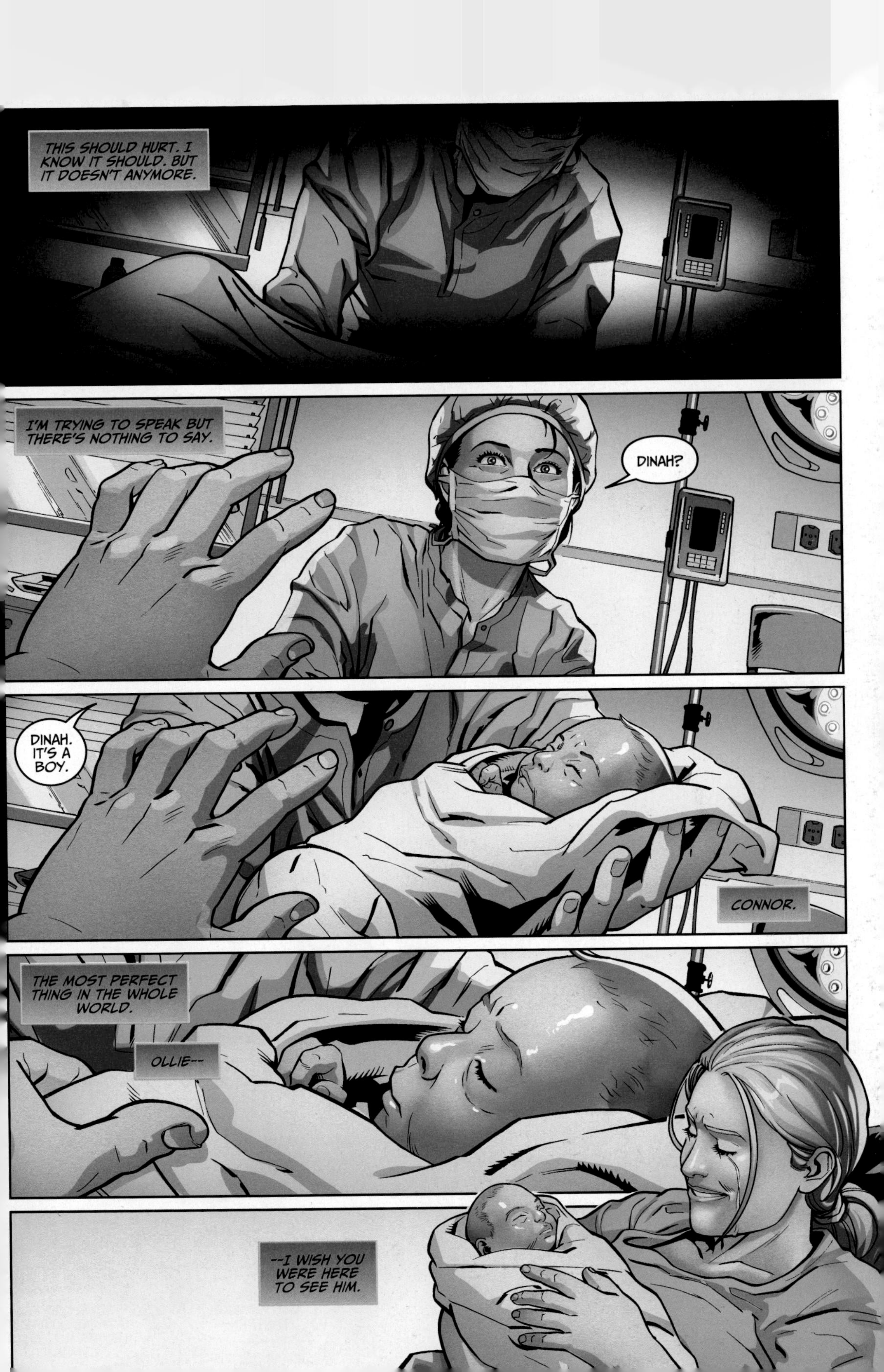
THIS SHOULD HURT. I KNOW IT SHOULD. BUT IT DOESN'T ANYMORE.
I'M TRYING TO SPEAK BUT THERE'S NOTHING TO SAY.
DINAH?
DINAH. IT'S A BOY.
CONNOR.
THE MOST PERFECT THING IN THE WHOLE WORLD.
OLLIE--
--I WISH YOU WERE HERE TO SEE HIM.

"PRIORITY ALERT! ALL MEMBERS!"
THE WATCHTOWER.
WHAT IS IT?
I DON'T KNOW. BUT IT LOOKS LIKE--
"-- A PERSON!"
CONTACT SUPERMAN!!
HE'S ON THEMYSCIRA WITH WONDER WOMAN. HE'S NOT RESPONDING!
USE THE WATCHTOWER LASER. TAKE IT OUT!
IT'S MOVING TOO FAST.
WHERE ARE THE GREEN LANTERNS?
JOHN AND HAL ARE INBOUND BUT THEY'RE NOT GOIN TO MAKE IT!

FLASH. GET
O CHICAGO!
"THERE WILL
BE WOUNDED
WHO NEED US.
"DAMN IT!
IMPACT IN
3..."
KSSSH
KSSSHH
KSSSHHH
"2..."
NO.

DESPERO!!
I WON'T LET YOU HARM THESE PEOPLE, MONSTER. HOW DARE YOU ATTACK THE EARTH!
HOW DARE I--?
I DON'T KNOW WHAT SICK GAME YOU'RE PLAYING HERE, SINESTRO. BUT YOU KNOW FULL WELL WHY I'M HERE.
YOUR DAMN CORPS ATTACKED ME AND THREW ME DOWN HERE TO--
STOP IT, DESPERO! I DON'T WANT TO HURT YOU.

WHAT ARE YOU TALKING ABOUT?
WHAT ARE YOU DOING?
YOU DON'T HAVE TO... JUST STAND DOWN!
WHAT IS THIS?!
RELEASE MY HANDS!!
SINESTRO!
PLEASE. DON'T MAKE ME DO THIS!

CRRK
HAL. I... I DIDN'T HAVE A CHOICE.
HE WOULD HAVE...
I UNDERSTAND, SINESTRO. WE SAW.
I'M JUST GLAD YOU ARRIVED IN TIME TO STOP HIM FROM DOING MORE DAMAGE.
THERE WILL BE PEOPLE WHO NEED US IN THAT BUILDING.
I DIDN'T THINK THERE COULD BE ANY REDEMPTION FOR SINESTRO.
BUT--ALL OF THE GOOD HE'S DONE THESE LAST MONTHS--I STILL CAN'T BELIEVE IT.
I THINK WE NEED TO BELIEVE IT, JOHN--

"--PEOPLE CAN CHANGE."
OH, MY GOD!
WHAT ARE YOU DOING HERE, HARLEY?
I TOLD THEM I WAS YOUR SISTER. HOSPITALS ARE SO TRUSTING.
HE'S INCREDIBLE.
I HAVE PRESENTS!
HERE. A BOW AND ARROW SET-- NEVER TOO YOUNG FOR THAT.
A FAKE MUSTACHE. THAT OBVIOUSLY NEEDS NO EXPLANATION.
AND, HERE...
WHAT IS THAT?
IT'S A MUZZLE!

WHY WOULD HE NEED A MUZZLE?
PLEASE. I CAN'T TELL YOU HOW USEFUL MUZZLES WERE WHEN I FIRST GOT MY HYENAS.
HE'S A BABY. NOT A PACK HUNTER.
YEAH. BUT BABIES SCREAM. AND WHEN *YOU* SCREAM, CRAZY, DEAFENING, SMASHY POWER COMES OUT. IF HE GETS THAT POWER EARLY...
GIVE ME THE MUZZLE.
UM... DINAH.
YOUR BAG IS GLOWING GREEN. IS THAT A THING IT DOES?
NO.
NOT NOW.

DINAH.
AND NOW YOUR GLOWING BAG IS TALKING.
YOU CAN HEAR THAT TOO, RIGHT?
IS IT SAFE TO SPEAK?
YES. IT'S OKAY, GUY.
IT'S TIME, DINAH.
HOW LONG?
OUR FULL FORCE HAS GATHERED BUT WE'LL NEED THREE WEEKS TO MOVE OUR SECRET WEAPON.

WE ONLY GET ONE SHOT AT THIS. WE NEED EVERYONE WE HAVE. WE'LL NEED YOU AND YOUR RESISTANCE FORCE READY ON THE GROUND WHEN WE COME.
I UNDERSTAND.
THE GREEN LANTERN CORPS, THE SINESTRO CORPS, SUPERMAN'S FORCES.
ARE YOU SURE YOU CAN FIGHT THIS WAR WITHOUT THE PLANET BEING TORN APART?
NO--
WHICH IS WHY WE'RE BRINGING OUR *OWN* PLANET.
WAR IS COMING

ONLY TWO SENTRIES.
THIS IS THE SINESTRO CORPS WE'RE TALKING ABOUT. THEY'RE NOT EXACTLY THE MOST DISCIPLINED OUTFIT. I'M SURPRISED SINESTRO COULD TALK ANY OF THEM INTO STAYING OUT IN THE MIDDLE OF SPACE FOR THIS LONG.
YOU'RE SUPPOSED TO BE THE BEST SCOUT WE HAVE, BAGLEE. HOW DID THEY SEE YOU?!
YOU'RE RIGHT IN FRONT OF ME AND I CAN BARELY SEE YOU!
THEY DIDN'T SEE ME! YOU BROUGHT MOGO!
ONE DAY MORE

OKAY. WE HAVE TWO YELLOW LANTERNS FLEEING FAST. UNFORTUNATELY, THEY WILL HAVE ALERTED SINESTRO TO OUR PRESENCE ALREADY, SO THERE'S NO NEED TO BE SUBTLE.
THE GUARDIANS HAVE AUTHORIZED DEADLY FORCE AGAINST THE SINESTRO CORPS.
"I'D PREFER IT DIDN'T COME TO THAT BUT THESE GUYS HAVE KILLED A LOT OF US.
"SO, I SURE AS HELL WON'T JUDGE YOU IF YOU WANT TO BE JUST A LITTLE BIT MORE SADISTIC THAN USUAL WHEN DEALING WITH THEM.
"IF YOU WANT TO AIM FOR PARTICULARLY TENDER AREAS, GO FOR IT."
YOU SHOULD CONTACT OUR ALLIES ON EARTH IMMEDIATELY. LET THEM KNOW WE'RE SPEEDING UP OUR TIMELINE. SURPRISE IS NO LONGER AN OPTION.
NO, IT'S NOT. SINESTRO KNOWS WE'RE COMING NOW.

PEOPLE OF EARTH.
A WAR IS COMING TOMORROW.
WE DID NOT START THIS WAR. WE DID NOT *INVITE* THIS WAR. BUT JR WORLD WILL SOON BE ATTACKED BY AGENTS WORKING FOR POWERFUL INTERGALACTIC BEINGS.
THESE SELF-PROCLAIMED *"GUARDIANS OF THE UNIVERSE"* SEEK TO UNDO EVERYTHING WE'VE ACHIEVED.
"THE ROGUE NATIONS WE'VE TOPPLED.
"THE WARS WE'VE ENDED.
"THE PRIVATE FORCES THAT WE HAVE ESTABLISHED AS SUPER-POWERED PEACE-KEEPERS ACROSS THE GLOBE.
ALL OF IT WILL BE FOR NOTHING. IF THESE GUARDIANS AND THEIR GREEN LANTERN CORPS SUCCEED, EVERYTHING WE'VE FOUGHT FOR WILL BE LOST.
YES, WE HAVE POWERFUL ALLIES--

--BUT I URGE EVERY SINGLE PERSON ON EARTH TO JOIN US IN OPPOSING THESE ALIEN INVADERS.
FIGHT FOR YOUR WORLD. DO NOT LET THESE ALIENS TAKE YOUR WAY OF LIFE FROM YOU--FROM YOUR FAMILIES.
GET OUT ON THE STREETS AND FIGHT FOR YOUR FUTURE!
IS HE...?
YES. HE'S ELOQUENTLY AND PASSIONATELY INSPIRING THE ENTIRE WORLD TO GET OUT AND BE HUMAN SHIELDS.
THE GREEN LANTERNS' ATTACK WILL BE SEVERELY HAMPERED IF THERE ARE CIVILIANS EVERY--
DEET
FILE RECEIVED
WHAT'S THAT?
THEY'RE CODES.
CODES TO WHAT?
THEY'RE CODES TO THE WATCHTOWER COMMUNICATIONS SYSTEM.
I THINK OUR MYSTERY FRIEND ON THE INSIDE JUST GAVE US THE MEANS TO KNOW THE JUSTICE LEAGUE'S MOVEMENTS BEFORE THEY MAKE THEM!

"TO CHILDREN."
MOMMY WILL COME BACK FOR YOU, CONNOR.
BLP.
I PROMISE.
"TO HUSBANDS--
"--AND WIVES."
RENEE, I...
IT'S OKAY, KATE. I'LL SEE YOU OUT THERE.
YOU NEED TO STAY BEHIND. I COULDN'T HANDLE IT IF ANYTHING WERE TO HAPPEN TO YOU, LITTLE MUSTACHE.
TOMORROW, THE TYRANNY OF SUPERMAN WILL BE ENDED BY ALL OF YOU.
AND I'M SORRY I CAN'T FIGHT ALONGSIDE YOU.

ZATANNA. I BELIEVE IT'S TIME. IF YOU'D BE SO KIND TO SEND MS. KYLE, MRS. LANCE AND I TO THE BATCAVE...?
ALFRED?
PLEASE, MASTER BRUCE. DID YOU REALLY THINK I'D SIT BACK WHILE OUR FRIENDS WENT TO WAR?
NO... I GUESS NOT.
TROPSNART OT EHT EVACTAB.
ZATANNA. IF THE UNTHINKABLE HAPPENS, DON'T GET INVOLVED. WE WILL NEED YOU FOR PLAN "B."
NOW, SEND ME TO ORACLE.
THEY THINK YOU'R STAYING HERE, SAFE BRUCE.
THEY THINK WHAT I NEEDED THE TO THINK. ORACLE HAS AGREED NOT TO SAY ANYTHING.
I DON'T CARE HOW BROKEN I AM.
I WILL MAKE SURE SUPERMAN GOES DOWN.

"Absolute Freaking Carnage" Mike S. Miller Artist
"Ground Assault" Bruno Redondo Julien Hugonnard-Bert Artists
Rex Lokus Colorist Cover Art by Stephane Roux

WHEN THE SKY TURNS GREEN?
THAT'S THE MESSAGE GUY SENT. WHEN THE SKY TURNS GREEN, WE BEGIN THE GROUND ASSAULT.
HE SAID THERE'S NO WAY WE'RE GOING TO MISS IT.
ALL OF OUR PEOPLE ARE IN PLACE?
THANKS TO OUR MYSTERY PERSON ON THE INSIDE, I'M PATCHED INTO THE JUSTICE LEAGUE COMMUNICATIONS SYSTEM. WE'LL KNOW EVERY MOVE THEY MAKE.
BLACK CANARY AND ALFRED ARE IN YOUR CAVE, WAITING FOR THE GO.
CATWOMAN, HUNTRESS AND BATWOMAN ARE IN THE FIELD ALONG WITH MY FATHER AND HIS GCPD RESISTANCE FIGHTERS. HARLEY QUINN IS OUT THERE, TOO.
SHE SHOULDN'T BE WITH US. SHE'S A CRIMINAL AND A LIABILITY.
IF BLACK CANARY SAYS SHE FIGHTS, THAT'S THE END OF THE DISCUSSION.
YOU'RE NOT HERE TO JUDGE PEOPLE, BRUCE. YOU'RE HERE FOR YOUR TACTICAL MIND.
AND WHEN CATWOMAN HEARS I LET YOU OUT OF THE TOWER, SHE'S GOING TO KICK MY...
WHAT IS IT?

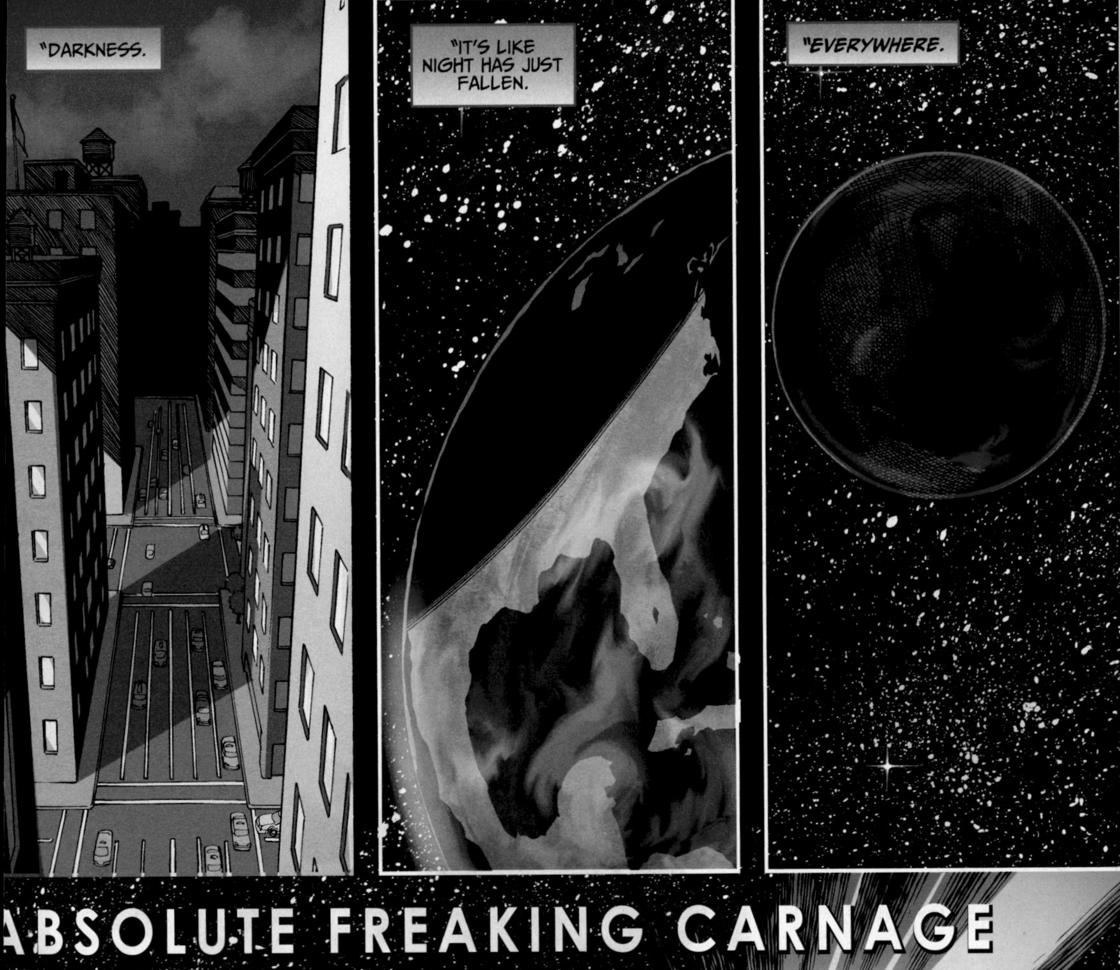
"DARKNESS.
"IT'S LIKE NIGHT HAS JUST FALLEN.
"EVERYWHERE.
ABSOLUTE FREAKING CARNAGE

"THE GREEN LANTERNS ARE HERE--

"--AND THEY'VE BROUGHT THEIR BIGGEST CORPS MEMBER."
THEY BROUGHT MOGO?
AND WHAT LOOKS LIKE HALF THE CORPS.
HAL. JOHN.
SO... THIS IS ALL KINDS OF NUTS, RIGHT?
OH. HEY, LOOK--

"--YOUR SAVIORS' ARE HERE.
HERE WE GO.
SO, BEFORE THIS WHOLE INTERGALACTIC WAR THING, I JUST WANTED TO MAKE ONE LAST APPEAL FOR SANITY.
HELP US TAKE DOWN THE SINESTRO CORPS. THEN STAND TRIAL. YOU GET JUDGED FAIRLY. THE EARTH KEEPS SPINNING. NONE OF OUR FRIENDS DIE.
NO ONE DIES IF YOU LEAVE.

DON'T DO THIS, MAN. HAVE SOME PERSPECTIVE. YOU HAVE SUPER-VISION OR WHATEVER. TAKE A GOOD LOOK AT YOUR 'ALLIES.'
THEY'RE THE SINESTRO CORPS. CHOSEN BECAUSE THEY CAN INSTILL FEAR. THEY'RE THE VERY DEFINITION OF BAD GUY. HALF OF THEM ARE JUST SPIKES AND SHARP TEETH.
THAT GUY'S NAME IS ARKILLO.
HE LITERALLY HAS 'KILL' IN HIS NAME.
FOR NOW, THEY ARE USEFUL. THE ENEMY OF MY ENEMY--
IS SOMETIMES JUST A MUCH WORSE ENEMY.
YOU'RE NOT GOING TO WIN THIS. I DON'T WANT TO HURT YOU--ANY OF YOU. I'M TRYING TO PROTECT YOU, AND OUR FRIENDS.
WE DON'T NEED YOUR 'PROTECTION.' WE KNOW HOW YOUR 'GUARDIANS' HAVE GUARDED WORLDS IN THE PAST.
WE CAN STOP THIS, CLARK.
YOU AND ME. JUST WORDS.
YOU GET THAT, RIGHT?
GUY...

YOU WERE ALWAYS AN EMBARRASSMENT TO THE JUSTICE LEAGUE.
NOW, YOU'RE AN EMBARRASSMENT TO YOUR PLANET, TOO.
OH, SCREW YOU, YOU SANCTIMONIOUS PIECE OF--
CRK
HNGG!
I DON'T WANT YOU IN THIS FIGHT.

ATTACK!
GUY!
JOHN. TELL HAL TO GET OUT OF THERE!
WHAT?
IT'S A FEINT! THEY'RE NOT REALLY ATTACKING!
"THE GUARDIANS WANT THE SINESTRO CORPS DESTROYED.
"YOU THINK WE CAME ALL THE WAY OUT HERE WITHOUT A PLAN?
"MOGO. A LIVING PLANET WITH A WILL-POWERED WEAPON.
"WHY DO YOU THINK WE DRAGGED HIM ACROSS SPACE?
"HOW MUCH WILL DO YOU THINK SOMETHING THAT SIZE HAS?"

MY GOD...
NO.
FREAKING.
WAY.
WHAT IS THAT?
GUY SAID WE WOULDN'T MISS IT.
NAAAAW.
PRETTY.

DINAH. THE SKY IS GREEN. THAT'S OUR SIGNAL.
BATMAN? YOU'RE SUPPOSED TO BE SAFE IN THE TOWER OF FATE.
NOT NOW. DO YOU HAVE THE WEAPON?
I DO.
I'M AFRAID YOU ONLY GET ONE SHOT, MISS LANCE.
I KNOW OU'RE ANGRY. NOW WHAT HE TOOK FROM YOU.
BUT WHEN THE CHANCE COMES TO TAKE HIM DOWN... TRY NOT TO KILL HIM.
REMEMBER WHO CLARK WAS.
I REMEMBER.
CLCK
BUT I CAN'T GUARANTEE ANYTHING--
--THE GUY WITH THE BEST AIM IS DEAD.

SPACE.
CYBORG! WHAT'S HAPPENING?
THE WATCHTOWER CAMERAS ARE SHOWING SOME SORT OF MASSIVE BLAST. I CAN'T SEE OUR PEOPLE.
HAL, JOHN, SUPERMAN, SINESTRO. I'VE LOST THEM ALL.
SUPERMAN!

SUPERMAN, ARE YOU THERE?!
CYBORG.
SUPERMAN! ARE YOU OKAY? ARE ALL OF OUR PEOPLE--?
"A QUARTER OF THE SINESTRO CORPS WERE JUST WIPED OUT."
CYBORG. TELL ALL OF OUR FORCES TO PULL BACK INTO EARTH'S ATMOSPHERE.
THEY CAN'T HIT US LIKE THAT AGAIN WITH THE PLANET BEHIND US.

ORACLE TO ALL RESISTANCE. MOGO TOOK OUT A QUARTER OF THE SINESTRO CORPS.
WE'VE THINNED THEM BUT THEY'RE BRINGING THE WAR BACK TO EARTH LIKE WE THOUGHT THEY WOULD.
ARE WE GO?
YES. THEY'RE STILL IN THERE. THEY'RE WAITING FOR INSTRUCTIONS AND STAYING CLOSE TO THE JUSTICE LEAGUE TELEPORTERS.
TAKE YOUR PILLS AND BE READY. WE'RE TRYING TO TAKE OUT ONE OF THE SMARTEST PEOPLE ON EARTH AND A MAN WHO CAN CROSS CONTINENTS IN A HEARTBEAT.
THE GUARDS OUTSIDE AND EVERYONE INSIDE WILL BE AS INVULNERABLE AS WE ARE. DON'T HOLD BACK.
IF FLASH ISN'T DEALT WITH, OUR GROUND ATTACK COULD BE OVER BEFORE ANY OF US EVEN REALIZE. NOW'S THE ONLY CHANCE WE HAVE TO TAKE HIM OFF THE BOARD.
QUINN. DON'T DEVIATE FROM THE PLAN. JUST OPEN THE DOOR.
HI! I'M THE FIRST WAVE.
WHAT?
LOOK!

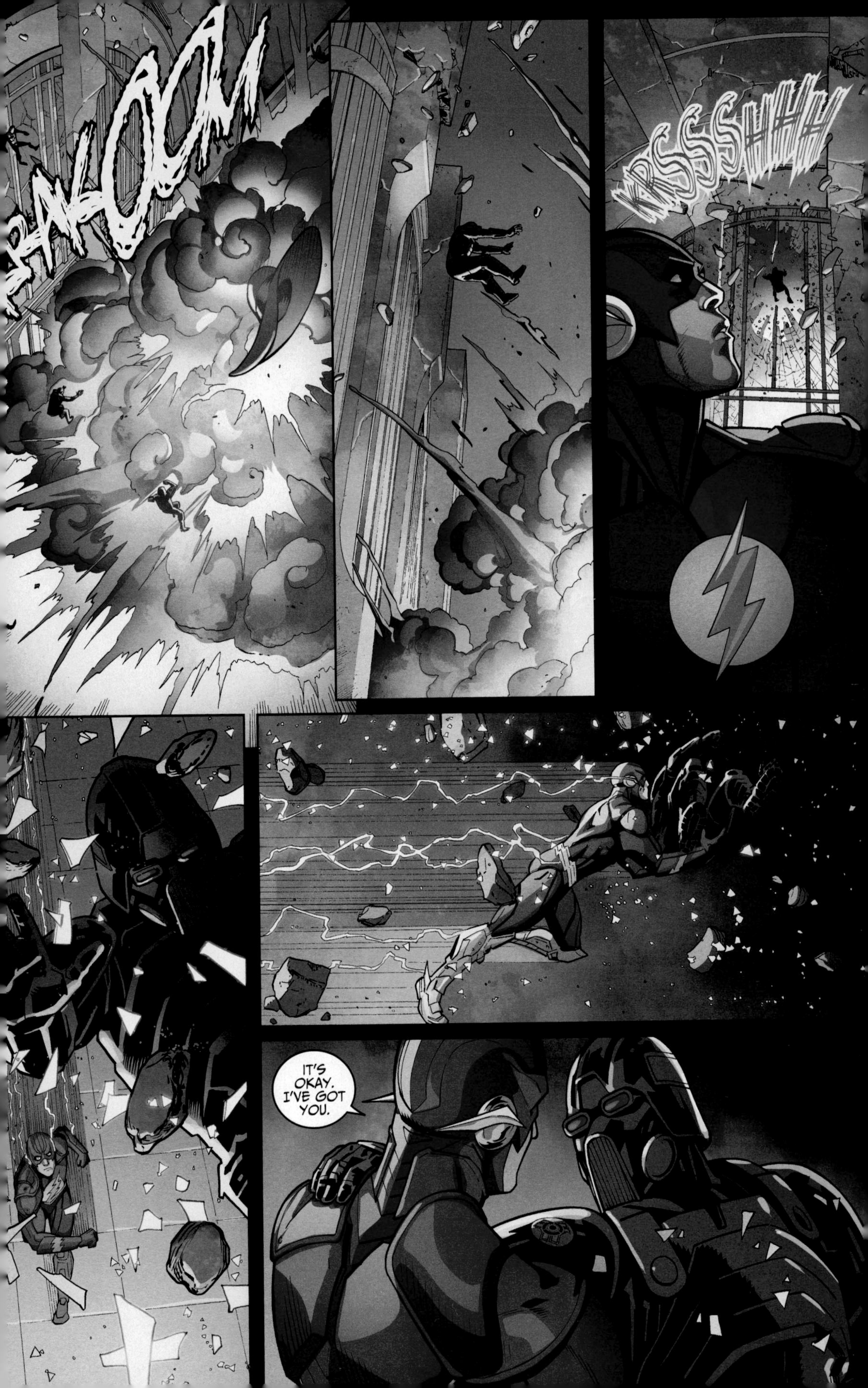
IT'S OKAY. I'VE GOT YOU.

MY HERO.
THIS ISN'T PERMANENT. BATMAN DOESN'T WANT YOU HURT.
BUT YOUR PART IN ALL THIS IS OVER.
CRCK

GO!
KOOOM
KOOM
FIND LUTHOR. FIND CYBORG. BRING THEM IN.

CHOOM
CYBORG, YOU CAN'T BE TAKEN. WE NEED YOU TO COORDINATE THE BATTLE. GET TO THE TELEPORTER. GET TO THE WATCHTOWER!
YOU'RE SURE?
GO. I'LL HOLD THEM OFF.
LEX, THERE'S SOMETHING...
SOMETHING AT THE EDGES. I CAN FEEL SOMEONE.
WHAT?
THEY'VE HACKED INTO OUR COMMUNICATIONS SYSTEM!
HOW?
I DON'T KNOW.
SHUT DOWN THE COMMUNICATIONS!
THD
IF I DO THAT, OUR FORCES WILL BE BLIND OUT THERE. AND, WHOEVER THIS IS, THEY CAN'T KNOW WE KNOW THEY'RE LISTENING.
DON'T WORRY. I'LL TRACE THEM. I WILL FIND THEM.
NO! SHUT IT DOWN!
CYBORG! SHUT IT DOWN!

WHAT IS THAT? WHAT DID YOU DO TO FLASH?!
I JUST STOPPED THE ELECTRICAL IMPULSES RUNNING DOWN HIS SPINE. DON'T WORRY. HE'S STILL THE FASTEST MAN ON EARTH FROM THE NECK UP.
MY FATHER MADE THIS DEVICE, DIDN'T HE? HE WAS ALWAYS THINKING OF CREATIVELY SICK NEW WAYS TO HURT HEROES.
YOU THINK HE HASN'T DEVELOPED SOMETHING THAT EXPLOITS ALL OF YOUR WEAKNESSES?
YOU HONESTLY THINK HE LOVES YOU, CATWOMAN? YOU THINK HE HASN'T CREATED SOMETHING THAT TEARS YOU OPEN? THAT EXPOSES YOU AND LEAVES YOU A BROKEN HUSK?
URGH. SHUT. UP.
YOU KNOW WHAT? ALL THE OTHER ROBINS WERE GOOD FOR BATMAN.
BUT YOU, DAMIAN--YOUR FATHER IS SO MUCH BETTER OFF WITHOUT YOU.
YOU'RE JUST THIS JOYLESS, ENTITLED, MOODY, WHINING LITTLE ****.
AND YOU HURT BRUCE.
I'M GOING TO PUNISH YOU FOR THAT.

HOLD HIM!
LOOK WHAT I FOUND! THE CREEPS HAD MY OLD COSTUME ON DISPLAY... WHICH SEEMS A BIT ICKY.
LUCKY THOUGH, HUH? 'CAUSE MY OTHER CLOTHES GOT BLOWN UP.
OOH. ONCE HE'S STOPPED STRUGGLING, CAN I PLEASE RUB HIS SHINY HEAD?
WE'VE TAKEN THE HALL OF JUSTICE. CYBORG ESCAPED BUT FLASH AND ROBIN ARE DOWN.
I SURRENDER. COMMISSIONER GORDON, I NEED TO SPEAK!
I HAVE TO TALK TO YOU.
IT'S ABOUT ORACLE.
LET HIM UP.

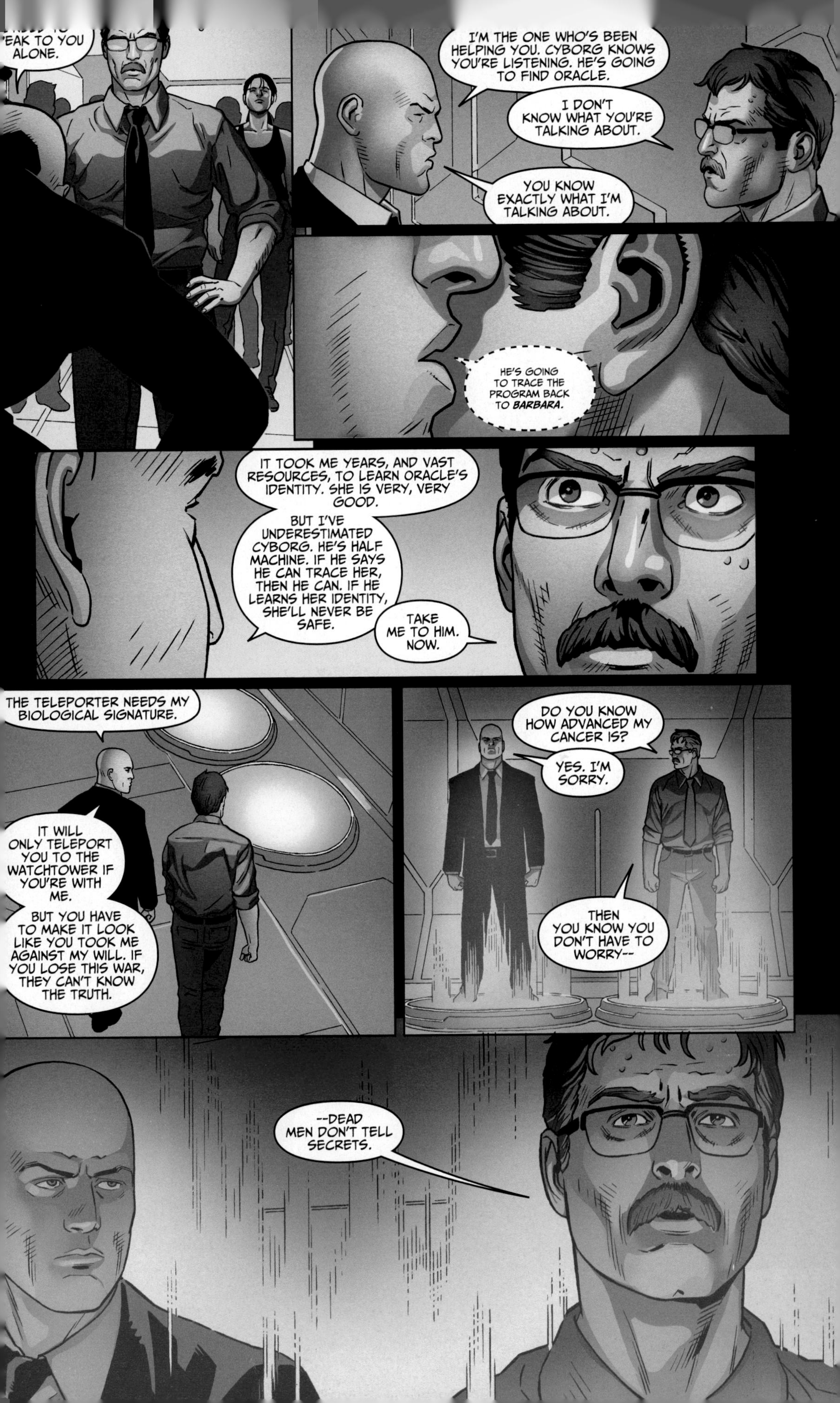

EAK TO YOU ALONE.
I'M THE ONE WHO'S BEEN HELPING YOU. CYBORG KNOWS YOU'RE LISTENING. HE'S GOING TO FIND ORACLE.
I DON'T KNOW WHAT YOU'RE TALKING ABOUT.
YOU KNOW EXACTLY WHAT I'M TALKING ABOUT.
HE'S GOING TO TRACE THE PROGRAM BACK TO BARBARA.
IT TOOK ME YEARS, AND VAST RESOURCES, TO LEARN ORACLE'S IDENTITY. SHE IS VERY, VERY GOOD.
BUT I'VE UNDERESTIMATED CYBORG. HE'S HALF MACHINE. IF HE SAYS HE CAN TRACE HER, THEN HE CAN. IF HE LEARNS HER IDENTITY, SHE'LL NEVER BE SAFE.
TAKE ME TO HIM. NOW.
THE TELEPORTER NEEDS MY BIOLOGICAL SIGNATURE.
IT WILL ONLY TELEPORT YOU TO THE WATCHTOWER IF YOU'RE WITH ME.
BUT YOU HAVE TO MAKE IT LOOK LIKE YOU TOOK ME AGAINST MY WILL. IF YOU LOSE THIS WAR, THEY CAN'T KNOW THE TRUTH.
DO YOU KNOW HOW ADVANCED MY CANCER IS?
YES. I'M SORRY.
THEN YOU KNOW YOU DON'T HAVE TO WORRY--
--DEAD MEN DON'T TELL SECRETS.

"Air Assault" **Mike S. Miller** Artist
"Gordon" **Bruno Redondo** **Julien Hugonnard-Bert** Artists
Rex Lokus Colorist Cover Art by **Jheremy Raapack & David Lopez** and **Santi Casas of Ikari Studio**

AIR ASSAULT
ORACLE. E'VE TAKEN HE HALL OF JUSTICE.
TELL HER I EXPLODED!
FLASH AND ROBIN ARE DOWN.
WHAT ABOUT CYBORG AND LUTHOR?
CYBORG ESCAPED, COMMISSIONER GORDON WAS INTERROGATING UTHOR, BUT, WELL... WE DON'T REALLY KNOW WHAT HAPPENED.
BULLOCK SAYS--
--GORDON TOOK LUTHOR INTO A TELEPORTER AND THEY BOTH WENT AWAY.
JIM WILL HAVE A REASON. WE HAVE TO TRUST YOUR FATHER.
WE'VE DONE WHAT WE CAN ON THE GROUND FOR NOW. SUPERMAN AND THE SINESTRO CORPS WILL BE FIGHTING MOST OF THIS WAR IN THE AIR. IT'S TIME TO JOIN THEM UP THERE.

"HUNTRESS, BATWOMAN. YOUR TEAM IS CLEARED.
"TAKE THE PLANES."
FERRIS AIRCRAFT
HAL!
I'M A LITTLE BUSY, CAROL!
HAL, THE RESISTANCE IS HERE!
THEY'RE ATTACKING FERRIS AIRCRAFT!
WHERE ARE YOU GOING?

ORACLE. I'M AWAY.
CAROL. ARE YOU OKAY?
I'M FINE. WHAT ARE THEY DOING HERE?
THEY'RE TRYING TO HELP.
LET THE PLANE GO.
THAT'S NOT GOING TO HAPPEN, GUY.
ARE WE REALLY GOING TO DO THIS?
NO. YOU'RE ABSOLUTELY NOT.
I CAN ACCEPT A CERTAIN AMOUNT OF MACHO CRAP FROM YOU TWO BUT I'M NOT GOING TO WATCH YOU BEAT EACH OTHER--

OKAY. IF THAT'S HOW IT'S GONNA BE.
YOU KNOW, I DID EVERYTHING I COULD TO CONVINC THEM. I TOLD THEM THERE WAS NO WAY T HAL JORDAN WOULD SIDE WITH FREAKING SINESTRO. I PROMISE THEM YOU'D COME AROUND.
HAL...?
IT'S OKAY. THEY'RE ON OUR SIDE.

YOU WERE SUPPOSED TO BE THE BEST OF US, MAN. YOU WERE THE ONE THE GUARDIANS CHOSE. ME, JOHN, KYLE...WE ONLY GOT THE GIG BECAUSE YOU WEREN'T AROUND.
YOU'RE NOT SUPPOSED TO BE SUPERMAN'S PUPPET.
YOU'RE SUPPOSED TO BE BETTER THAN THIS!
SHUT HIM UP.
I'M A CYNICAL BASTARD BUT I STILL HAD FAITH IN YOU, HAL.
END HIM!
YEAH. THAT'S NOT GOING TO WORK. THIS IS A WAR. YOU THINK WE DIDN'T PREPARE? WE DIDN'T COME ALL THIS WAY TO HAVE OUR ASSES HANDED TO US.
I NEVER STOPPED BELIEVING IN YOU, HAL--

--BUT THE GUARDIANS GAVE UP.
GANTHET?
FWOOOM

GUARDIAN!
NO!
I'VE GOT YOU!
I'D BETTER GET YOU OUT OF HERE. THIS PLACE IS PROBABLY GOING TO BE ENTIRELY COLLATERAL DAMAGE SOON.
I HOPE YOUR COMPANY IS INSURED FOR INTERGALACTIC WAR.

GUY!
GIVE HER TO ME!
IT'S OKAY, HAL. GUY SAVED ME. HE--
GIVE HER TO ME!!
CALM THE HELL DOWN!
YOU THINK THE GUARDIANS WILL LET YOU STAND AGAINST THEM WITH THEIR POWER?
GSSHH
HNNNG!
HAL JORDAN.
YOU HAVE BROKEN YOUR OATH.

YOU ARE NOT WORTHY OF WIELDING GREEN LANTERN'S LIGHT.
NO.
LET HIM GO!
SINESTRO!!
SAVE CAROL.
NO.
PLEASE!
WE NEED YOU IN THIS WAR.
SAVE HER YOURSELF.

ISN'T A MORAL QUANDARY!
JORDAN. CAROL WILL DIE.
TAKE IT. SAVE HER.
NOW--
"--FEEL HER FEAR.
"USE IT!"
THEY TRIED TO CAST YOU ASIDE. THEY TRIED TO TAKE HER FROM YOU.
THEY TORE US APART BUT THEY FORCED US TOGETHER AGAIN, HAL. YOU AND I. WE ARE ONE CORPS AGAIN.
LET US SHOW THEM FEAR.

DAMN IT! WHAT ARE THEY DOING?
ORACLE! TELL THE GREEN LANTERNS TO COME BACK TOGETHER.
"SUPERMAN AND THE SINESTRO CORPS ARE DELIBERATELY STRETCHING THEM OUT.
CRCK
"THEY'RE PREADING THE VAR ACROSS THE WORLD.
"THEY'RE MOVING THE BATTLE TO POPULATED AREAS.
"THEY KNOW THE GREEN LANTERNS WON'T TAKE OUT CIVILIANS.

WHAT IS IT? WHAT'S WRONG?
I DON'T KNOW.
THERE WAS SOMETHING IN THE SYSTEM AND THEN IT WAS GONE.
TIK TAK
THERE.
SOMEONE'S TRACING US!
WHAT?
THE REGIME KNOWS WE'RE LISTENING!
SHUT DOWN THE SERVERS!
IT'S TOO LATE!
THEY'RE COMING. THEY'RE MOVING THROUGH MY PROXIES LIKE THEY'RE NOT EVEN THERE!
CYBORG...

THE WATCHTOWER.
I KNOW YOU'RE THERE.
Biological signature recognized.
Lex Luthor teleporting to Watchtower.
LEX. I'VE TRACED THE HACK IN OUR COMMUNICATIONS TO SOMEONE CALLED ORACLE. WHOEVER IT IS THEY'VE MASKED THEMSELVES WELL BUT I'LL HAVE THEIR LOCATION IN--
CLCK
NO--
--YOU WON'T.
STOP THE TRACE AND COME QUIETLY.
WE'RE ON A SPACE STATION. THERE'S NO WAY YOU'D FIRE A WEAPON IN--

BANG
TING
YOU WANT TO MAKE ANY OTHER ASSUMPTIONS?
STOP THE TRACE OR THE NEXT BULLET HITS THE SOFT SIDE OF YOUR FACE.
I DON'T THINK YOU'RE GOING TO KILL ME. NOT IN COLD BLOOD. YOU'RE POLICE. I DON'T THINK YOU CAN.
DON'T DO ANYTHING STUPID.
I DON'T HAVE TO DO ANYTHING. I JUST HAVE TO WAIT.

YOUR VITALS ARE DROPPING.
YOU'RE DYING, AREN'T YOU?
DON'T FIGHT. IT'S OVER. THIS 'ORACLE' IS PUTTING UP A FIGHT BUT THE PROXIES ARE FALLING. I'LL HAVE THEIR LOCATION SOON.
I'M SORRY. YOU'RE OUT OF TIME.
IS THERE ANYTHING I CAN DO FOR YOU?
CHANNNN
NO!
I WON'T LET YOU FIND HER!

KNK
I DON'T UNDERSTAND TECHNOLOGY.
WHAT ARE YOU DOING?!!
BUT I KNOW NOTHING RUNS--
HNGGG!!
--IF YOU TURN IT OFF!

TNK
ZZSST
IT'S GONE.
THE TRACE HAS GONE!
DID THEY FIND US?
THEY MADE IT AS FAR AS GOTHAM AND THEN...I DON'T KNOW WHAT HAPPENED. IT JUST WENT AWAY.

COMMISSIONER GORDON? YOU DID IT. SHE'S SAFE.
IS HE...?
HE'S ALIVE BUT HE'S OUT OF ACTION.
YOU UNDERSTAND WHY I COULDN'T ACT DIRECTLY?
OF COURSE. IF WE FAIL, YOU WILL BE NEEDED ON THE INSIDE. YOU'RE SECRET IS SAFE, LUTHOR.
YOU DON'T LOOK SO GOOD.
NO.
DID YOU KNOW?
WHAT?
THAT THE PILLS INCREASED THE STRENGTH OF THE CANCER?
I KNEW. BUT WHAT CHOICE DID I HAVE?
COULD YOU TELEPORT CYBORG BACK DOWN TO THE WATCHTOWER FOR COLLECTION? AND GIVE ME A MOMENT TO MYSELF?
OF COURSE.
ORACLE?
COMMISSIONER?
I'M ON THE WATCHTOWER. CYBORG HAS BEEN SHUT DOWN. TELL BULLOCK AND MONTOYA THAT I'M TELEPORTING HIM BACK DOWN TO THEM IN THE HALL OF JUSTICE.
AND WHAT ABOUT YOU, COMMISSIONER GORDON?
COULD WE SPEAK ON A CLOSED CHANNEL?

DAD?
CLOSED CHANNEL - COMMISSIONER GORDON
BARBARA.
I WISH YOU COULD SEE THIS.
IT LOOKS SO BEAUTIFUL FROM UP HERE.
YOU CAN'T SEE ANY FEAR OR VIOLENCE OR HATRED. IT'S JUST... CALM.
SUPERMAN IS A CRIMINAL. NOTHING MORE. HE CAN'T BE ALLOWED TO HOLD ALL OF THIS IN HIS HANDS.
WE WON'T LET HIM, JIM.
IT'S GOOD TO HEAR YOUR VOICE, BATMAN.
IT'S BRUCE.
NO, IT ISN'T.
IT WAS NEVER 'BRUCE.'
NO.
IT'S BEEN AN HONOR TO SERVE GOTHAM WITH YOU.
KEEP SERVING HER.

DAD, BULLOCK IS STANDING BY. JUST GET TO THE TELEPORTER AND--
DAD? PLEASE... DON'T.
I'M SORRY, BARBARA.
BUT YOU'RE STRONG. YOU ALWAYS HAVE BEEN. STAY STRONG. THE RESISTANCE NEEDS YOU.
I LOVE YOU. I'M SORRY I HAVE TO SAY GOODBYE.
BATMAN. I GUESS I... I GUESS I GET TO DISAPPEAR ON YOU FOR ONCE.
JIM?
"JIM...?"

"Crashing to Earth" Mike S. Miller Artist J. Nanjan (NS Studios) Colorist
"Canary's Revenge" Tom Derenick Artist Rex Lokus Colorist
Cover Art by **Jheremy Raapack & David Lopez** and **Santi Casas of Ikari Studio**

HIGH ABOVE ERRIS AIRCRAFT BASE.
AAAAGHHH!
IT'S OKAY--
--I'VE GOT YOU, CAROL.
HAL?
WHAT HAPPENED TO YOU?

HE HAS FALLEN.
FALLEN? FALLEN FROM WHAT?
HE HAS FOUND A TRUER PATH, YOU SELF-RIGHTEOUS--
YOU WERE NEVER GOOD ENOUGH FOR THE CORPS.
WE ARE BETTER THAN YOUR CORPS!
PRIDE, SINESTRO. PRIDE WAS ALWAYS YOUR UNDOING.
PLUS YOUR HABIT OF KILLING INNOCENT PEOPLE. THAT UNDID A LOT.
YOU WERE--BOTH OF YOU--TERRIBLE MISTAKES.

FERRIS AIRCRAFT
GO. GO. GO!
HUNTRESS? I THOUGHT YOU ALREADY FLEW A PLANE OUT OF HERE!
I DID BUT IT GOT USED AS A BLUDGEONING WEAPON IN A FIGHT BETWEEN SINESTRO AND A GUARDIAN.
HUH?
COME ON. WHILE THEY'RE STILL FIGHTING EACH OTHER.
ORACLE--
--WE'RE AWAY!
BARBARA?
IT'S OKAY. I CAN TAKE OVER, IF YOU NEED ME TO--
NO.

THE BATCAVE
ORACLE TO BLACK CANARY. THE OTHERS ARE IN THE AIR. ARE YOU READY?
I'M READY, ORACLE.
OPENING THE HANGAR, MISS LANCE.
IF YOU'D BE SO KIND AS TO HIT SUPERMAN WITH A MISSILE FOR ME, I'D APPRECIATE IT.
OF COURSE, ALFRED.
CANARY. BATMAN WOULD LIKE A WORD.
WHAT IS IT?
OU'LL NEED TO BE VERY CAREFUL. THE PLANE IS AD-LINED, SO SUPERMAN N'T BE ABLE TO SEE IN."
AND?
WHY DO I NEED TO BE MORE CAREFUL IF SUPERMAN CAN'T SEE INTO YOUR PLANE?
HE MIGHT THINK YOU'RE ME.

GANTHET?
YES. IT IS TIME TO GO, GUY GARDNER. WE ARE NEEDED ELSEWHERE.
WHERE ARE YOU GOING?
TO DEAL WITH SOMEONE MORE IMPORTANT THAN YOU.
THERE IS NO ONE MORE IMPORTANT THAN ME!!
I HAVE WATCHED CIVILIZATIONS RISE AND FALL.
YOU ARE A SPECK IN HISTORY, SINESTRO.
YOU ARE ENTIRELY INSIGNIFICANT.
I HAVE NO MORE TIME FOR YOU.

NAAARGH!
YOU HAVE DONE ENOUGH DAMAGE, SUPERMAN.
WE WILL NOT ALLOW YOU TO DO MORE.

SINESTRO SAYS YOU'RE TO LIVE.
TARGET THE SINESTRO CORPS. WE WANT SUPERMAN ALONE.
FIRE!
KOOM
KOOM

WHO...?
RAK
KOOOOM
BRUCE.

CHZZZZZZZZ
KOOOM
EEEEEEEEEEEEEEE
HNG.

EEEEEEEEEEEEEEE
EEEEEEEEEEEE
EEEEEEEEEEEEE
EEEEEEEEEEEE
TOOOM
AGHHHHH!

IS HE DOWN?
CANARY. IS SUPERMAN DOWN?
NO.
I NEVER WANTED TO HURT YOU, DINAH. I NEVER WANTED TO HURT ANYONE.
OH, SHUT UP!
YOU WANT ME TO FEEL SORRY FOR YOU? YOU'RE THE MOST POWERFUL THING ON EARTH. STOP PRETENDING THAT SOMEONE ELSE IS TO BLAME. IT'S PATHETIC.
YOU HAVE SUPER-VISION. YOU'RE TELLING ME YOU CAN'T SEE THE EVIL, MURDERING DICTATOR YOU'VE BECOME?
I'M JUST GLAD LOIS DIED BEFORE SHE SAW YOU LIKE THIS.
YOU DON'T TALK ABOUT LOIS.
DO YOU REMEMBER WHAT I TOLD YOU AT OLLIE'S FUNERAL?

KRYPTONITE BULLET.
YOU WEREN'T FASTER THAN THAT. WERE YOU?

THD

HNG.
I TOLD YOU I'D TEAR YOU DOWN.

I MISSED ANYTHING VITAL ON PURPOSE.
I PROMISED BATMAN I'D GRANT YOU THE MERCY YOU DENIED OLLIE, YOU SON OF A BITCH.

EVERYTHING YOU'VE DONE TO HIM--

--AND BRUCE STILL CARES ABOUT YOUR WORTHLESS LIFE.

YOUR GN IS VER.
"AND THE EVIL YOU BROUGHT TO EARTH IS DONE.
"WITHOUT YOU, THE SINESTRO CORPS WILL CRUMBLE.
NAAARGHH--
"WITHOUT YOU, THEY WILL FALL APART."

"AND ALL THOSE YOU SOLD YOUR TWISTED DREAM TO--
"--ALL THOSE YOU TRICKED INTO FIGHTING FOR YOU--
SHAZ--
SHHH.
"--THEY ARE ALREADY FALLING."

THE HALL OF JUSTICE HAS FALLEN.
"WE ARE TAKING YOUR ALLIES.
"WE'RE HIDING THEM.
IS THAT SCOWL ESPECIALLY FOR THIS SITUATION, DAMIAN?
IT'S HARD TO TELL. IT LOOKS LIKE YOUR REGULAR, EVERYDAY SCOWL.
"THEY WON'T BE COMING TO RESCUE YOU."
STAY CALM. YOU'RE JUST INCAPACITATED. IT'S NOT PERMANENT.

OOH. I LIKE YOUR LITTLE LIGHTNING-BOLT EARS.
DO YOU HAVE THOSE BECAUSE YOU HEAR AT SUPER-SPEED?
DOOOO IIIII SOOOOUUUND SLLLOOOOWW TOOOOO YOOOUUUU?
IT'S OKAY. THEY TELL ME YOU'RE A NICE GUY WHO'S MADE A WHOLE LOT OF BAD DECISIONS.
BUT YOU'RE JUST A BIT BRAINWASHED. IT HAPPENS. BELIEVE ME. I'VE BEEN THERE.
YOU GET THIS GUY WHO'S ALL STRONG AND POWERFUL AND CONFIDENT, AND YOU JUST STARE INTO HIS DREAMY FACE AND DO WHAT HE SAYS, RIGHT?
I UNDERSTAND. BUT WE'RE GOING TO GET YOU THROUGH THIS. I PROMISE.
THEY'VE ASKED ME TO REPROGRAM YOU!
DON'T WORRY. I'M A PROFESSIONAL.
I HOPE YOU LIKE ELECTRO-SHOCK THERAPY AND PUPPETS!

GUY. CAN YOU HEAR ME?
CAN YOU SPEAK UP? I'M IN THE MIDDLE OF AN INTERGALACTIC WAR.
WE DID IT.
SUPERMAN IS DOWN.
WELL DONE, DINAH.
TELL GANTHET TO COME AND GET HIM.
GET THIS EVIL BASTARD OFF OF OUR PLANET.

KAL-EL OF THE PLANET KRYPTON.
YOU HAVE THE ABILITY TO INSTI GREAT FEAR.
NO...
HNNNG!
HNNGH.

CHZZZZZZ
CRASHING TO EARTH
WELCOME TO THE SINESTRO CORPS.

"Fall of the Gods" **Bruno Redondo** **Xermanico** **Julien Hugonnard-Bert** Artists **Rex Lokus** Colorist
"World's End" **Bruno Redondo** **Xermanico** **Julien Hugonnard-Bert** **Juan Albarran** Artists
J. Nanjan (NS Studios) Colorist Cover Art by **Jheremy Raapack & David Lopez** and **Santi Casas of Ikari Studio**

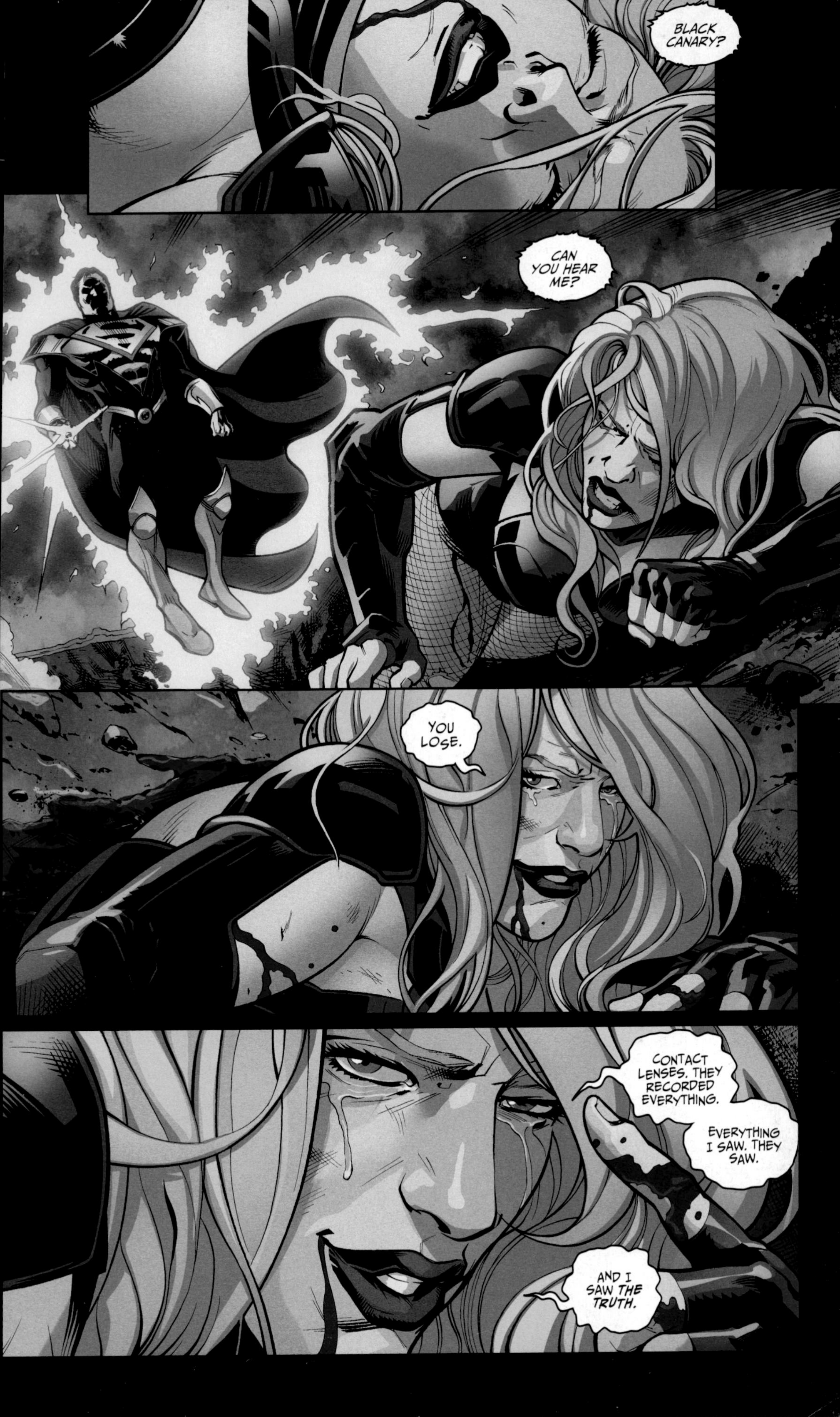
BLACK CANARY?
CAN YOU HEAR ME?
YOU LOSE.
CONTACT LENSES. THEY RECORDED EVERYTHING.
EVERYTHING I SAW. THEY SAW.
AND I SAW THE TRUTH.

DINAH...?
YOU'R BEING STREAM AROUND WORLD
YOU DON'T GET TO HIDE BEHIND YOUR HERO FAÇADE ANY LONGER.
YOU MIGHT WIN THIS WAR. YOU MIGHT KILL ALL THOSE WHO TRIED TO STOP YOU.
ATTACKS
BUT EVERYONE WILL...KNOW WHAT YOU ARE NOW.
THEY'LL KNOW YOU'RE... YOU'RE NOT THEIR SAVIOR.
THE WHOLE WORLD... WILL FEAR YOU.

CHZZZZZZ
IS THIS WHAT YOU WANT, GANTHET?
GRCK
ARE YOU HAPPY TO SEE SO MANY DIE?
LOOK AT ALL OF THIS!
"DO THE GUARDIANS CARE AT ALL?!"

STOP!
STOP FIGHTING!!
STEWART!
WHAT ARE YOU DOING?!
I DON'T--
YOU CAN'T STOP THIS, JOHN.
THE GUARDIANS DON'T CARE ABOUT YOUR WORLD.
SUPERMAN DOES.
WE NEED YOU ON OUR SIDE.

STOP REACTING.
CLOSE YOUR EYES AND THINK.
YOU'RE AN ARCHITECT.
SEE THE DESIGN. SEE WHAT THE GUARDIANS HAVE PLANNED FOR YOUR WORLD. SEE HOW THEY WANT THIS UNIVERSE BUILT AROUND THEM.
CAN YOU SEE IT?
I DON'T WANT TO...
I KNOW--

SHNK
--THE TRUTH IS PAINFUL.
HRK...
IT'S ALL RIGHT, JOHN STEWART. YOU WILL HELP US AFTER ALL.

JORDAN!!
JOHN!
QUICKLY. YOU NEED TO GET HIM MEDICAL ATTENTION.
THE WAR...
DON'T WORRY ABOUT THE WAR!
SAVE YOUR FRIEND.
LANTERN OF SECTOR 2814--
NO.
--DECEASED. SCANNING FOR SUITABLE REPLACEMENT.

WHO DID THIS?
WHO DID THIS??!
IT DOESN'T MATTER. THIS IS NO TIME FOR A PERSONAL VENDETTA. THE WORLD NEEDS YOU.
IT WAS GUY GARDNER.
YOU KNOW, DON'T YOU?
GARDNER ALWAYS FELT UNWORTHY. HE WAS ONLY YOUR REPLACEMENT. YOU KNOW HE ALWAYS WANTED TO BE THE ONLY ONE.
THE ONLY GREEN LANTERN OF THIS SECTOR.
STEWART WAS TRYING TO PROTECT BOTH SIDES. HE WAS TRYING TO STOP THE DEATH.
HE DIDN'T EVEN SEE HIM COMING. I'M SORRY.
I KNOW IT HURTS TO HEAR BUT I BELIEVE HE WANTED THIS.
HE BROUGHT THIS WAR HERE. HE PLANNED ALL OF THIS.

GUY!!!
HAL! DON'T--
SSHHHH
HRAARGH!!
YOU BROUGHT THEM HERE!!
YOU DID THIS!
NO. I DIDN'T WANT ANY OF THIS.
HAL. PLEASE. I'M NOT GOING TO FIGHT YOU.

HAL. DON'T LISTEN TO HIM, MAN.
HE WILL TAKE WHAT IS YOURS. YOUR WORLD. YOUR LOVED ONES.
THIS ISN'T YOU. THIS IS THE RING. FIGHT IT!
DO IT!!
HAL. YOU CAN STILL BE THE BEST OF US.
AAAARGH!!

MY GOD.
SHUT DOWN THE FEED!
WE WANT PEOPLE TO SEE--
SHUT IT DOWN! I NEED TO TALK TO OUR PEOPLE.
"I AM ORDERING A FULL RETREAT.
"THE GREEN LANTERNS ARE FALLING.
"GUY GARDNER IS DOWN.
"BLACK CANARY...
"BLACK CANARY HAS FALLEN."
IT'S OVER. WE CAN'T WIN THIS.
GANTHET COULD STILL--
NO.

"SUPERMAN IS WEARING A RING POWERED BY FEAR."
YOU WERE TO BE TAKEN TO OA TO STAND TRIAL.
HOWEVER, THAT IS NO LONGER REQUIRED.
"WE JUST STREAMED IMAGE OF HIM MURDERING SOMEO ALL OVER THE GLOBE."
YOU ARE A THREAT, NOT JUST TO THIS WORLD, BUT TO WORLDS BEYOND THIS ONE.
GANTHET! DON'T DO THIS!
"HE'LL BE FUELLED BY THE FEAR OF A BILLION PEOPLE."
YOUR EXISTENCE CANNOT CONTINUE.
"HE'LL BE UNSTOPPABLE."

HE TRIED TO EXECUTE YOU, SUPERMAN.
I KNOW.
MOGO! FIRE!!
AAAARGHH!!

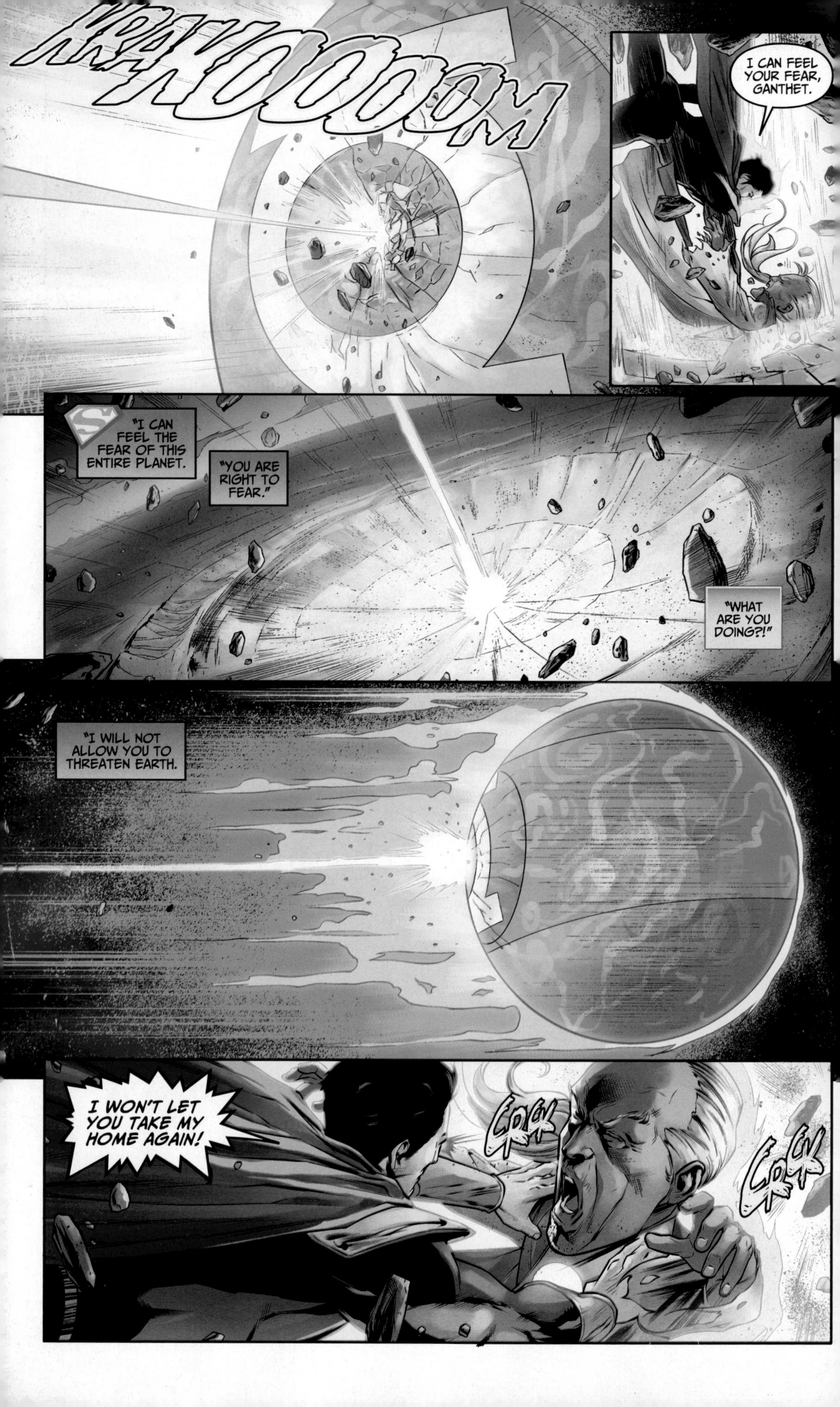
KRAKOOOOOM
I CAN FEEL YOUR FEAR, GANTHET.
"I CAN FEEL THE FEAR OF THIS ENTIRE PLANET.
"YOU ARE RIGHT TO FEAR."
"WHAT ARE YOU DOING?!"
"I WILL NOT ALLOW YOU TO THREATEN EARTH.
I WON'T LET YOU TAKE MY HOME AGAIN!
CRACK
CRACK

YOU CAN'T DO THIS!!
WE DON'T NEED YOU, "GUARDIAN."

"WE CAN GUARD OURSELVES."
IT IS DONE.
WHAT HAVE WE DONE?
WHAT WE SET OUT TO DO.
"WE SAVED THE WORLD."
END

EPILOGUE
**** FATE.

"WHERE...?"
WHERE AM I?
YOU ARE IN ANOTHER PLACE. ANOTHER UNIVERSE. ANOTHER EARTH. YOU WERE FATED TO DIE.
I... INTERVENED.
WAAAAH!
CONNOR?
YOU BROUGHT HIM, TOO?
OF COURSE.
WHO--?

OLLIE?
I DON'T UNDERSTAND...
THE OLIVER QUEEN OF THIS EARTH LOST HIS DINAH LANCE FIVE YEARS AGO.
THE TWO OF YOU HAVE NEVER MET.
I CAN'T GUARANTEE YOU HAPPINESS, BUT I CAN GIVE YOU EACH OTHER.
BUT MY WORLD... SUPERMAN.
I WON'T TAKE YOU BACK THERE.
I HAVE ALREADY BROKEN TOO MANY RULES.
AND YOU HAVE SACRIFICED ENOUGH.
GOODBYE, CONNOR LANCE-QUEEN. I HOPE YOU HAVE A WONDERFUL LIFE.

CONNOR...?
YES.
I'VE ALWAYS LIKED THAT NAME.
I KNOW.
HELLO, PRETTY BIRD.

encils from Injustice: Year Two #10 by Bruno Redondo